How We Survived
Communism and Even
Laughed

Slavenka Drakulić

How We Survived Communism and Even Laughed

HarperPerennial

A Division of HarperCollinsPublishers

"A Chat with My Censor" was previously printed in the March 12, 1988, issue of *The Nation*. "My First Midnight Mass" was previously printed in the February 12, 1990, issue of *The Nation*.

A hardcover edition of this book was published in 1992 by W. W. Norton and Company, Inc. It is here reprinted by arrangement with W. W. Norton and Company, Inc.

First HarperPerennial edition published 1993.

Library of Congress Cataloging-in-Publication Data

Drakulić, Slavenka, 1949–
    How we survived communism and even laughed / Slavenka Drakulić.—
1st HarperPerennial ed.
    p.   cm.
Originally published: [New York]: W.W. Norton & Co.
ISBN 0-06-097540-7 (pbk.)
1. Communism—Europe, Eastern. 2. Europe, Eastern—Politics and government—1945– .
3. Drakulić, Slavenka, 1949– . 4. Authors, Yugoslav—Biography. I. Title.
HX365.5.A6D73   1993
335.43'0947—dc20   92-54849

---

04 05 RRD H 20

CONTENTS

Acknowledgments

Books come in life like children do. First there is a seed, then it grows. The idea for this book now seems too obvious: there were so many articles and books written on Eastern Europe after 1989, but none of them spoke about women, their lives, their feelings. My traveling on assignment for *Ms. Magazine* to Hungary, Poland, Czechoslovakia, Bulgaria and East Germany in January and February of 1990 made me realize that. And even if I had been previously to these countries, that trip made me see what we all went through in the last forty-five years. It was as if only the present could unlock the door to the past.

Yet, this book is not only on women. To understand their (our) situation, one has to see the system at work behind.

I am grateful to Robin Morgan, who sent me on this trip in the first place. To Gloria Steinem, for trusting me more than I trust myself. To Mary Cunnane, my editor at Norton, who gave me not only the opportunity but also the support to finish this book. To Kate Mosse, my editor at Hutchinson, for writing cheerful letters in bleak times. To my friends Andrea, Alemka, Jasmina, Vesna, Dorina and Vanja here in Zagreb, who find time to help me with their comments. And to my daughter Rujana who, as always, bravely sustained my unbearable changes of temper.

Perhaps I should first thank the women in the countries I visited and who wholeheartedly gave me their help and their time, even if I didn't know them. Instead of naming them all – because it is impossible – I dedicate this book to all women of Eastern Europe who, too, made possible the changes in 1989.

Epigraph

We are the needy relatives, we are the aborigines, we are the ones left behind – the backward, the stunted, the misshapen, the down-and-out, the moochers, parasites, conmen, suckers. Sentimental, old-fashioned, childish, uninformed, troubled, melodramatic, devious, unpredictable, negligent. The ones who don't answer letters, the ones who miss the great opportunity, the hard drinkers, the babblers, the porch-sitters, the deadline-missers, the promise-breakers, the braggarts, the immature, the monstrous, the undisciplined, the easily offended, the ones who insult each other to death but cannot break off relations. We are the maladjusted, the complainers intoxicated by failure.

We are irritating, excessive, depressing, somehow unlucky. People are accustomed to slight us. We are cheap labor; merchandise may be had from us at a lower price; people bring us their old newspapers as a gift. Letters from us come sloppily typed, unnecessarily detailed. People smile at us, pityingly, as long as we do not suddenly become unpleasant.

As long as we do not say anything strange, sharp; as long as we do not stare at our nails and bare our teeth; as long as we do not become wild and cynical.

– György Konrád, *To Cave Explorers from the West*

Introduction

The Trivial Is Political

The title of my book feels wrong, I kept thinking as my plane soared off the runway at Zagreb airport. We have not yet survived communism, and there is nothing to laugh about.

I took off for London on Thursday, 27 June 1991, the day after the two secessionist republics of Slovenia and Croatia declared their independence from the rest of Yugoslavia, one year after holding their first free elections since World War II. The previous night Slovenia's borders with Italy and Austria were closed in anticipation of clashes between its territorial defense forces and the Yugoslavian federal army commanded by Serbian generals in Belgrade. That morning I called the airport repeatedly: 'Will the Adria Airways flight to London be leaving this morning?'

'We don't know. Who knows? Who can tell?' said the young woman at the other end. I could sense panic in her voice, the same panic I recognized in myself, growing like a strange little fruit in the pit of my stomach.

On the news that evening in London I heard about the first deaths in Slovenia. There was a war going on, a real war, not in some far off country about which we knew little, but right in the heart of Europe, close to the home I had left behind. I watched the TV bulletins reporting helicopters being shot out of the sky, tanks being blown up, and people being killed. This was not how any of us had imagined the future eighteen months before, when the great wave of change had swept across Eastern Europe. This was not how it was supposed to be. I felt cheated.

Suddenly I caught myself thinking about fruit and about how nothing had changed. We had thought that after the revolution peaches would be different – bigger, sweeter, more golden. But as I stood in line at a stall in the street market I noticed that the peaches were just as green, small, and bullet-hard, somehow pre-revolutionary. The tomatoes were still far too expensive. The strawberries – still sour, the oranges – still dry and wrinkled. Then I thought about dust, the fine yellow dust that used to cover the shop windows and windowsills, the buildings and the cars, as if nobody cared. That hadn't changed either.

What did change were the faces of the politicians on TV, the names of the major streets and squares, the flags, national anthems, and monuments. These days in Duke Jelačić Square in Zagreb, the Croatian capital, sun glitters on the refurbished baroque facades, and the newly-erected bronze statue of the duke casts a long shadow across the marble pavement. He holds his sword aloft pointing the way forward, but the more I look at him, the more I am afraid he is really still directing us into the past, as if forty-five years of living under communism cannot be erased from our collective consciousness without a substitution. There was no duke when the place was called Republic Square. It was in the spring of 1990 that the newly elected democratic government erected a statue of Duke Jelačić to replace the one that had been destroyed by Tito's partisans soon after World War II as a 'relic of the past' with no place in their 'new communist society'. It felt as though the new democracies of Eastern Europe were so weak, so fearful of the legacy of their communist past, that they had to abolish the symbols of the old regime with the same eagerness – violence even – as the revolutionary governments had used after 1945.

I clearly remember when it all began. Just before he retired, a journalist colleague returned from the Austria-Hungary border in mid-September 1989, crying with excitement. 'East Germans are crossing the border by the thousand. I didn't think I would ever live to see this!' Neither did I. That is how you are trained in this part of the world, not to believe that change is possible. You are trained to fear change, so that when change eventually begins to take place, you are suspicious, afraid, because every change you ever experienced was always for the worse. I remember that my own first reaction to my colleague's news, besides happiness, was fear, as if I were experiencing an earthquake. Much as I desired the collapse of the old system, the ground was shaking beneath my feet. The world I had thought of as permanent, stable, and secure was suddenly falling apart all around me. It was not a pleasant experience.

Later on events began unfolding with such speed that enthusiastic reporters and theoreticians – who, by the way, were taken just as much by surprise as the common people – barely had time to develop fully an Eastern European domino theory or to decide whether authentic revolutions were taking place. This disorientation mixed with hope followed me in January 1990 when I traveled through Hungary, Czechoslovakia, Poland, East Germany, and Bulgaria. I knew that, just like every other trip I had ever made to these countries, it would feel very much like revisiting my own past – the shortages, the distinctive odors, the shabby clothing. After all, we had all long suffered under the same ideology. Despite this, I needed to see with my own eyes what was going on.

One of the first things I noticed on these travels was the influence of Hollywood movies on our media and, consequently, on our way of thinking. In newspapers and on TV revolutions looked spectacular: cut barbed wire, seas of lighted candles, masses chanting in the streets, convulsive embraces and tears of happiness, people chiseling pieces from the Berlin Wall. A famous Hollywood director once said that movies are the same as life with the boring parts cut out. I found that this was precisely right. The boring parts of the revolutions had simply finished up on the floors of television studio cutting rooms all over the world. What the world had seen and heard were only the most dramatic and symbolic images. This was all right, but it was not all. Life, for the most part, is trivial.

These trivial aspects, the small everyday things, were precisely what I wanted to see: how people ate and dressed and talked, where they lived. Could they buy detergent? Why was there so much rubbish all over the streets? In short, I wanted to take all these fragments of recent reality, as well as my own memories of life in a communist country, and sew them back together.

Growing up in Eastern Europe you learn very young that politics is not an abstract concept, but a powerful force influencing people's everyday lives. It was this relationship between political authority and the trivia of daily living, this view from below, that interested me most. And who should I find down there, most removed from the seats of political power, but women. The biggest burden of everyday life was carried by them. Even if they fully participated in revolutionary events, they were less active and less visible in the aftermath of those events.

After the revolutions women still didn't have time to be involved; they still distrusted politics. At the same time, they deluded themselves that the new democracies would give them the opportunity to stay at home and perhaps rest for a while. There was something else, too: somebody had to take responsibility for finding food and cooking meals, a task made no easier – indeed, in some countries made more difficult – by the political changeover.

Women's lives, by no means spectacular, banal in fact, say as much about politics as no end of theoretical political analysis. I sat in their kitchens – because that was always the warmest room in their poorly heated apartments – listening to their life stories, cooking with them, drinking coffee when they had any, talking about their children and their men, about how they hoped to buy a new refrigerator or a new stove or a new car. Then we would go shopping together or go to their workplaces or attend meetings or go walking through the streets, into restaurants, drugstores, churches, and beauty parlors. Even if, in some cases, we had never met before, I felt that the pattern of our lives was almost identical. We had all been forced to endure the same communist system, a system that ground up people's lives in a similar way wherever you lived; then, of course, as women, we shared a perspective on life that was different from men's. Ours was trivial, the 'view from below.' But trivia is political.

As the English historian Timothy Garton Ash wrote in his book *We, the People* concerning the changes in Eastern Europe: 'Sometimes a glance, a shrug, a chance remark will be more revealing than a hundred speeches.'

I understand that in the West today 'the end of communism' has become a stock phrase, a truism, a common expression supposed to indicate the current state of things in Eastern Europe. It sounds marvelous when you hear it in political speeches or read it in the newspapers. The reality is that communism persists in the way people behave, in the looks on their faces, in the way they think. Despite the free elections and the celebration of new democratic governments taking over in Prague, Budapest, and Bucharest, the truth is that the people still go home to small, crowded apartments, drive unreliable cars, worry about their sickly children, do boring jobs – if they are not unemployed – and eat poor quality food. Life has the same wearying immobility; it is something to be endured, not enjoyed. The end of communism is still remote because communism, more than a political ideology or a method of government, is a state of mind. Political power may change hands overnight, economic and social life may soon follow, but people's personalities, shaped by the communist regimes they lived under, are slower to change. Their characters have so deeply incorporated a particular set of values, a way of thinking and of perceiving the world, that exorcising this way of being will take an unforseeable length of time.

In Hungarian director Peter Basco's long-forbidden film, *Key Witness*, there is a scene in an amusement park where, in 'The Tunnel of Socialist Horror,' Marx, Engels, Lenin, and Stalin surge out of the darkness to frighten little children. In a way things were easier in pre-revolutionary Eastern Europe. We had only to enter the tunnel, blame everything – all our private, as well as public, woes – on the party. Eighteen months ago, as we finally left the tunnel, perhaps we found that things were not exactly as we dreamed they would be. Somehow we slowly realized that we had to create our own promised land, that from now on we would be responsible for our own lives, and that there would no longer be any convenient excuses with which to ease our troubled consciences. Democracy is not like an unexpected gift that comes without effort. It must be fought for. And that is what makes it so difficult.

At this precise moment, perhaps, the title of my book feels wrong. We may have survived communism, but we have not yet outlived it.

London, 4 July 1991

You Can't Drink Your Coffee Alone

She is dead. Her grave is covered with ivy and tiny blue forget-me-nots. A candle is burning; her mother must have been here recently. But I haven't been here – not once since the day she was buried, five years ago. It's not that I've forgotten her, just the opposite: I could not face it, her death, the absurdity of it. In August 1985, when she poisoned herself with the gas from the stove in her new apartment, she was thirty-six years old.

I don't know how to tell Tanja's story or even why it should be important. She wasn't a hero; sometimes I think she was a coward. The many fine threads that connected her to life simply unraveled one by one so that she chose death over life. Before she let the gas run, she sealed all the doors and windows with tape and washed the dishes. To this day, I am not sure whether I should resent her for this tidiness in preparing her own death or take it as a last sign of her wish to go on, to live. I see her standing over the kitchen sink, compulsively washing as if it is the most important thing to do at that moment, because it will postpone what is to come soon – the loneliness of death. When she died, it looked like it was one more unhappy love story. It was somehow easier for people not to think too much of it, not to look for the other, less obvious side of it. It was safer to reduce her suicide to a cliché.

That winter her lover died. He died during open heart surgery, but a surgeon, his friend, said he wouldn't survive anyway; his heart was too weak. 'Worn out,' this is the expression he used. They were both journalists at the same newspaper. He had a wife working at a public library and a teenage daughter, and wouldn't leave them. 'I admire him for his loyalty to his family,' Tanja used to say, and I could not tell whether she really meant it or whether cynicism was her way of dealing with this fact. The summer before his death, she got pregnant. With two divorces behind her and no children, she wanted his child badly. But it so happened that his wife was pregnant at the same time. He stopped seeing Tanja. Devastated, she didn't have the strength to have a baby all by herself and had an abortion. When he heard that she was no longer pregnant, he returned to her. She took him back. A few months later, his wife gave birth to a little girl, whom he adored.

The winter started badly. The gray smoggy air felt as if one was breathing in dirty cotton. She wrote an article that kicked up a lot of dust. It was against nationalizing all privately owned pinball machines on behalf of the state-owned lottery company. 'If today we take away pinballs, because we believe they are doing the work instead of their owners ... sometime soon we might nationalize privately

2

owned trucks, because they too are working instead of their owners. Or close private hairdressing shops – for what is hand combing or haircutting compared to the work done by one single electric hooded dryer, not to mention curlers, shampoos, conditioners, hair-spray, etc.' She cleverly and humorously compared the case of pinball machines with the case of a Soviet citizen, Vasili Mihailovich Pilipenko, and the polemics in the Soviet press that very summer about whether he could or couldn't keep the horse that he had found and reared. In Byelorussia, the law treated draft animals as an unlawful source of enrichment without work. To a foreigner, this article would look very innocent. What harm can writing about pinball machines do? But we had brought to perfection the social game called 'reading between the lines,' so of course it was understood that her article was not about pinball machines, but about the privatization of the economy. Yugoslavia has passed through different stages of economic reform. One of the stages was privatization, letting small businesses develop by private investment in order to heal the economy. But times were changing, the economic policy was taking a different turn. Read through ideological glasses, her article was clearly political. In fact, her political mistakes were severe. First of all, she took the 'capitalist' orientation of the state seriously, ie, allowing the development of private small business, and she defended it. Then she insulted and ridiculed the judicial system by showing that the parliament in the Socialist Republic of Croatia – as in every one-party state – is only a formality. Her article, naive as it seems today, speaking 'only' about pinball machines, revealed the functioning and hypocrisy of the communist state. She mocked it, and she had to be punished for that.

After a week of 'consultations' (an expression for talks with the party heads about the most recent instructions on editorial policy in newspapers or, in effect, unofficial censorship) the editorial board of her newspaper published a hundred-and-fifty-word boxed statement entitled 'Explanation from the editors' – as far as I remember, perhaps the last of its kind. It was not unusual, though. This practice was a hangover from the past, when editors were directly responsible to the party man, a censor, for articles. But it was also an efficient instrument for settling accounts with the 'enemies of the people,' in other words whoever wrote against (or rather, not according to) a party policy, so editors didn't give up the practice easily. Later on, it served for eliminating 'disobedient' journalists. In Tanja's case, it looked as if the editors were explaining to the public the severe 'error' that had occurred in the newspaper. But everyone knew that it was their token sackcloth and ashes, a declaration written for the party bosses, not for the public: 'The editorial board . . . has considered the opinions expressed in the article. After careful analysis and discussion, the board concluded that publishing the article represents a serious editorial mistake in the professional and political meaning of the word . . . Our newspapers do and will support legislative and other activities stimulating individual and private enterprise, except for those not in accordance with the activity of the Socialist Alliance of Working People and the ideological guidelines of the League of Communists.' However, this was the message they were trying to convey: 'We, the editorial board, admit our mistake in not having had such complete control of our newspaper, that unfortunately, the unwanted ideas appeared. We will make sure it doesn't happen again.'

I can see Tanja, sitting beside her desk on the seventh floor of the glass and aluminum building on Ljubljanska Avenija, reading the fresh newspaper that still smelled of printer's ink, leaving her hands black with it. She read it and read it, thinking, as had many victims before her: *No, this is not possible; this must be some terrible mistake.* Perhaps that was the moment when she finally saw through all the illusions surrounding her. She saw the glass wall of her reality splintering.

What struck her most? Not only the words, but the meaning of the action. The rejection of her as a journalist, as a colleague, as a person. Her editorial board, people she knew, people she had worked with for more than ten years, other party members – because she *was* a member of the Communist Party, she couldn't possibly have been a commentator on the most important daily newspaper without this pedigree – they all renounced her. As soon as the 'explanation' came out, she felt as if she had ceased to exist: 'You know,' she told me, 'my colleagues don't dare to say hello to me any more. I feel as if I'm invisible. Nobody wants to have coffee with me, but you can't drink your coffee alone.' This was it, the sudden invisibility. From then on, she was put 'on ice' – ignored, invisible, nonexistent, a non-journalist, a non-person. She could write and even get her monthly pay, but nothing got into print. The worst was that nobody could tell her how long this situation would last – because nobody could tell how long the system would last.

I don't know when she started to contemplate suicide. Was it after the abortion, when she was alone and badly hurt? Was it after her lover's death in March? And how much did the situation at her job help her to make up her mind? Perhaps it would be wrong to say that she killed herself just because of the psychological pressure she suffered after she wrote that article. But it seems to me that it would also be wrong to say that she died because of her lost lover. The earth beneath her was crumbling. She couldn't change her job, there was nowhere to go. She was aware that she would suffer the destiny of a dissident. In her life, she couldn't see anything to hold on to.

This was a time of desperate attempts by the Communist Party to maintain communism – or socialism, as we called it – in Yugoslavia. After Tito's death in 1980 there were predictions that the country would fall apart. This was a time of reform of the educational system, of attacks on artistic and intellectual independence (the *Bijela Knjiga*, index of 'unwanted' writers, artists, intellectuals), of political trials . . . It had become obvious that the system of 'self-management' Yugoslavia was so proud of was a ruse, invented to make you believe that you – not the government or the party – are to blame. It was the most perfect system among the one-party states, set up to internalize guilt, blame, failure, or fear, to teach you how you yourself should censor your thoughts and deeds and, at the same time, to make you feel that you had more freedom than anyone in Eastern Europe.

But Tanja was wrong in one thing: she believed it would go on forever like that – the same newspaper, the same faces, the same cold climate of fear and silent accusations, the immobility of the system – forever the same. What communism instilled in us was precisely this immobility, this absence of a future, the absence of a dream, of the possibility of imagining our lives differently. There was hardly a way to say to yourself: This is just temporary, it will pass, it must. On the contrary, we learned to think: This will go on forever, no matter what we do. We can't change it. It looked as if the omnipotent system had mastered time itself. For our generation, it seemed that communism was eternal, that we were sentenced to it and would die before seeing it collapse. We were not revolutionaries trying to destroy it, to bring it down. We were brought up with the idea that it is impossible to modify the system, to change it eventually from within. Still, if only Tanja had waited. One's life is not a waiting room in a provincial train station, where one sits waiting for a train that might never come. Yet a week before she died, she'd cut her hair short. I don't think women do this if they are thinking of dying. She struggled to survive but in the end she lost.

It was the last day of October when I finally visited her grave. The sun was already low and cold, but the yellow and red leaves of a climbing vine like one in my garden on the outside walls of the cemetery made the landscape somehow pleasant, and I didn't resist the thought of sitting down. I didn't try to escape. As I looked at the place where she was buried, I could feel where the memory of her dwelled in me, deep down, under the diaphragm, where I myself had secluded her, squeezing the last remembrance of her, of that summer day in August 1985, and turning it into a tight little ball of pain.

And, as if I was redeemed by this visit to her grave, that hidden memory, the ball in me started melting, and I could see her face, as she sat across from me in my room – I could hear her voice, as if we were both there, now. The day before she died, she had come to tell me something. A farewell, perhaps. I should have guessed by the way she was talking about suicide. But she had talked about it so many times, she had theorized about it as an act of will, an honorable way out in a desperate situation. The flat tone of her voice in the haze of a sleepy summer afternoon, music from somewhere sneaking through the heavy air, sun, heat, it all misled me. We drank beer. As the afternoon passed, she looked more and more as if she were drowning in the armchair, deeper and deeper, and I remember thinking that it was eating up her skinny, fragile body. Suddenly it seemed to me that her hands were trembling, that as she uttered the word 'death', she felt a chill running down her spine. But I was not sure. I blame myself for that.

I heard about her death early in the morning. The telephone rang, there was that voice saying incomprehensible words, words that I was supposed to understand. 'Do you hear me, do you hear me?' The voice repeated the question, meeting a silence on my end of the line. At first, I didn't respond, didn't say anything. Then a scream came. I heard it coming out of my throat, but at the same time I was another person, standing beside me. Looking from a distance at the person holding a telephone and screaming, totally mute, I was without feelings, numb.

Later that day, I went to Tanja's apartment. She had moved in recently, about a month ago, after her parents exchanged their three-room apartment for a smaller one. She had long waited for this to happen and decorated the apartment very carefully, picking out in antique shops by herself every porcelain cup, silver mirror, or vase. Her books were neatly arranged on the shelf, her typewriter open. On the table there was a fresh bunch of daisies. It was as if by tidying her small apartment, she had tried to confront the chaos in her life. On the surface it looked as if she had found the strength to go on. Inside, she just could not go on any more.

There was only one detail that made me cry: a Bible on her bed. It was upside down. I turned it over and took a quick look at the underlined text. It was about life after death. She was not a religious person; in fact, she was an atheist. But in the last moments of her life, nothing else was left to her, and she turned to the Bible. For months afterwards all I could think of was the bunch of fresh daisies and the Bible. I wonder if the people who wrote that 'explanation' ever think about her? Her death was wrong, it was useless, and only today can I see the full absurdity of it. Perhaps communism is collapsing, but what is the price? How many more victims like Tanja will it claim – not big heroes, political prisoners, or dissidents, but people who just couldn't stand it anymore?

Standing at her grave, I simply wish, for her sake, that there is another life after death. I imagine she is sitting there having a coffee with someone. There must be someone up there, because, as she said, you can't drink your coffee alone.

Pizza in Warsaw, Torte in Prague

We were hungry, so I said 'Let's have a pizza!' in the way you would think of it in, say, New York, or any West European city – meaning 'Let's go to a fast-food place and grab something to eat.' Jolanta, a small, blond, Polish translator of English, looked at me thoughtfully, as if I were confronting her with quite a serious task. 'There are only two such places,' she said in an apologetic tone of voice. Instantly, I was overwhelmed by the guilt of taking pizza in Poland for granted. 'Drop it,' I said. But she insisted on this pizza place. 'You must see it,' she said. 'It's so different from the other restaurants in Warsaw.'

We were lucky because we were admitted without reservations. This is a privately owned restaurant, one of the very few. We were also lucky because we could afford a pizza and beer here, which cost as much as dinner in a fancy hotel. The restaurant was a small, cozy place, with just two wooden tables and a few high stools at the bar – you couldn't squeeze more than twenty people in, even if you wanted to.

It was raining outside, a cold winter afternoon in War-saw. Once inside, everything was different: two waiters dressed in impeccable white shirts, with bow ties and red aprons, a bowl of fresh tropical fruit on the bar, linen nap-kins and the smell of pizza baked in a real charcoal-fired oven. Jolanta and I were listening to disco music, eating pizza, and drinking Tuborg beer from long, elegant glasses. Perhaps this is what you pay for, the feeling that you are somewhere else, in a different Warsaw, in a dreamland where there is everything – pizza, fruit juice, thick grilled steaks, salads – and the everyday life of shortages and poverty can't seep in, at least, for the moment.

Yet to understand just how different this place is, one has to see a 'normal' coffee shop, such as the one in the modernistic building of concrete and glass that we visited the same day. Inside neon lights flicker, casting a ghostly light on the aluminum tables and chairs covered with plas-tic. This place looks more like a bus terminal than like a *kawiarnia*. It's almost empty and the air is thick with ciga-rette smoke. A bleached blond waitress slowly approaches us with a very limited menu: tea, some alcoholic beverages, Coke, coffee. 'With milk?' I ask.

'No milk,' she shakes her head.

'Then, can I get a fruit juice perhaps?' I say, in the hopes of drinking just one in a Polish state-owned restaurant.

'No juice.' She shakes her head impatiently (at this point, of course, there is no 'sophisticated' question about the kind of juice one would perhaps prefer). I give up and get a cup of coffee. It's too sweet. Jolanta is drinking Coke, because there is Coke everywhere – in the middle of Warsaw as, I believe, in the middle of the desert. There may be neither milk nor water, but there is sure to be a bottle of Coke around. Nobody seems to mind the paradox that even though fruit grows throughout Poland, there is no fruit juice yet Coke is everywhere. But here Coke, like everything coming from America, is more of a symbol than a beverage.

To be reduced to having Coke and pizza offered not only as fancy food, but, what's more, as the idea of choice, strikes me as a form of imperialism, possibly only where there is really very little choice. Just across the street from the private restaurant, where Jolanta parked her tiny Polski Fiat, is a grocery store. It is closed in the afternoon, so says a handwritten note on the door. Through the dusty shop window we can see the half-empty shelves, with a few cans of beans, pasta, rice, cabbage, vinegar. A friend, a Yugoslav living in Warsaw, told me that some years ago vinegar and mustard were almost all you could find in the stores. At another point, my friend noticed that shelves were stocked with prune compote. One might easily conclude that this is what Poles probably like the best or why else would it be in stores in such quantities? But the reason was just the opposite: Prune compote was what they had, not what they liked. However, the word 'like' is not the best way to explain the food situation (or any situation) in Poland. Looking at a shop window where onions and garlic are two of the very few items on display, what meaning could the word 'like' possibly have?

Slowly, one realizes that not only is this a different

reality, but that words have a different meaning here, too. It makes you understand that the word 'like' implies not only choice but refinement, even indulgence, *savoir-vivre* – in fact, a whole different attitude toward food. It certainly doesn't imply that you stuff yourself with whatever you find at the farmer's market or in a grocery that day. Instead, it suggests a certain experience, a knowledge, a possibility of comparing quality and taste. Right after the overthrow of the Ceausescu government in Romania in December 1989, I read a report in the newspaper about life in Bucharest. There was a story about a man who ate the first banana in his life. He was an older man, a worker, and he said to a reporter shyly that he ate a whole banana, together with the skin, because he didn't know that he had to peel it. At first, I was moved by the isolation this man was forced to live in, by the fact that he never read or even heard what to do with a banana. But then something else caught my attention: '*It tasted good*,' he said. I can imagine this man, holding a sweet-smelling, ripe banana in his hand, curious and excited by it, as by a forbidden fruit. He holds it for a moment, then bites. It tastes strange but 'good.' It must have been good, even together with a bitter, tough skin, because it was something unachievable, an object of desire. It was not a banana that he was eating, but the promise, the hope of the future. So, he liked it no matter what its taste.

One of the things one is constantly reminded of in these parts is not to be thoughtless with food. I remember my mother telling me that I had to eat everything in front of me, because to throw away food would be a sin. Perhaps she had God on her mind, perhaps not. She experienced

World War II and ever since, like most of the people in Eastern Europe, she behaves as if it never ended. Maybe this is why they are never really surprised that even forty years afterwards there is a lack of sugar, oil, coffee, or flour. To be heedless – to behave as if you are somewhere else, where everything is easy to get – is a sin not against God, but against people. Here you have to think of food, because it has entirely diverse social meanings. To bring a cake for dessert when you are invited for a dinner – a common gesture in another, more affluent country – means you invested a great deal of energy to find it if you didn't make it yourself. And even if you did, finding eggs, milk, sugar, and butter took time and energy. That makes it precious in a very different way from if you had bought it in the pastry shop next door.

When Jaroslav picked me up at Prague airport, I wanted to buy a torte before we went to his house for dinner. It was seven o'clock in the evening and shops were already closed. Czechs work until five or six, which doesn't leave much time to shop. 'The old government didn't like people walking in the streets. It might cause them trouble,' said Jaroslav, half joking. 'Besides, there isn't much to buy anyway.' My desire to buy a torte after six o'clock appeared to be quite an extravagance, and it was clear that one couldn't make a habit of bringing a cake for dessert. In the Slavia Café there were no pastries at all, not to mention a torte. The best confectioner in Prague was closed, and in the Hotel Zlatá Husa restaurant a waitress repeated 'Torte?' after us as if we were in the wrong place. Then she shook her head. With every new place, my desire to buy a torte diminished. Perhaps it is not that there are no tortes – it's just hard to find them at that hour. At the end, we went to the only shop open until eight-thirty and bought ice cream. There were three kinds and Jaroslav picked vanilla, which is what his boys like the best.

On another occasion, in the Bulgarian capital Sofia, Evelina is preparing a party. I am helping her in the small kitchen of the decaying apartment that she shares with a student friend, because as an assistant professor at the university, she cannot afford to rent an apartment alone. I peel potatoes, perhaps six pounds of them. She will make a potato salad with onions. Then she will bake the rest of them in the oven and serve them with . . . actually nothing. She calls it 'a hundred-ways potato party' – sometimes humor is the only way to overcome depression. There are also four eggs for an omelet and two cans of sardines (imported from Yugoslavia), plus vodka and wine, and that's it, for the eight people she has invited.

We sit around her table: a Bulgarian theater director who lives in exile in Germany, three of Evelina's colleagues from the university, a historian friend and her husband, and the two of us. We eat potatoes with potatoes, drink vodka, discuss the first issue of the opposition paper *Demokratia*, the round-table talks between the Union of Democratic Forces and the communist government, and calculate how many votes the opposition will get in the forthcoming free elections – the first. Nobody seems to mind that there is no more food on the table – at least not as long as a passionate political discussion is going on. '*This* is our food,' says Evelina. 'We are used to swallowing politics with our meals.

For breakfast you eat elections, a parliament discussion comes for lunch, and at dinner you laugh at the evening news or get mad at the lies that the Communist Party is trying to sell, in spite of everything.' Perhaps these people can live almost without food – either because it's too expensive or because there is nothing to buy, or both – without books and information, but not without politics.

One might think that this is happening only now, when they have the first real chance to change something. Not so. This intimacy with political issues was a part of everyday life whether on the level of hatred, or mistrust, or gossip, or just plain resignation during Todor Živkov's communist government. In a totalitarian society, one *has* to relate to the power directly; there is no escape. Therefore, politics never becomes abstract. It remains a palpable, brutal force directing every aspect of our lives, from what we eat to how we live and where we work. Like a disease, a plague, an epidemic, it doesn't spare anybody. Paradoxically, this is precisely how a totalitarian state produces its enemies: politicized citizens. The 'velvet revolution' is the product not only of high politics, but of the consciousness of ordinary citizens, infected by politics.

Before you get here, you tend to forget newspaper pictures of people standing in line in front of shops. You think they serve as proof in the ideological battle, the proof that communism is failing. Or you take them as mere pictures, not reality. But once here, you cannot escape the *feeling* of shortages, even if you are not standing in line, even if you don't see them. In Prague, where people line up only for fruit, there was enough of all necessities, except for oranges or lemons, which were considered a 'luxury.' It is hard to predict what will be considered a luxury item because this depends on planning, production, and shortages. One time it might be fruit, as in Prague, or milk, as in Sofia. People get used to less and less of everything. In Albania, the monthly ration for a whole family is two pounds of meat, two pounds of cheese, ten pounds of flour, less than half a pound each of coffee and butter. Everywhere, the bottom line is bread. It means safety – because the lack of bread is where real fear begins. Whenever I read a headline 'No Bread' in the newspaper, I see a small, dark, almost empty bakery on Vladimir Zaimov Boulevard in Sofia, and I myself, even without reason, experience a genuine fear. It makes my bread unreal, too, and I feel as if I should grab it and eat it while it lasts.

Every mother in Bulgaria can point to where communism failed, from the failures of the planned economy (and the consequent lack of food, milk), to the lack of apartments, child-care facilities, clothes, disposable diapers, or toilet paper. The banality of everyday life is where it has really failed, rather than on the level of ideology. In another kitchen in Sofia, Ana, Katarine and I sit. Her one-year-old daughter is trying to grab our cups from the table. She looks healthy. 'She is fine now,' says Ana, 'but you should have seen her six months ago. There was no formula to buy and normal milk you can hardly get as it is. At one point our shops started to sell Humana, imported powdered milk from the dollar shops, because its shelf life was over. I didn't have a choice. I had to feed my baby with that milk, and she got very, very sick. By allowing this milk to be sold, our own government poisoned babies. It was even on TV; they had to put it on because so many babies in Sofia got sick. We are the Third World here.'

If communism didn't fail on bread or milk, it certainly failed on strawberries. When I flew to Warsaw from West Berlin, I bought cosmetics, oranges, chocolates, Nescafé, as a present for my friend Zofia – as if I were going home. I also bought a small basket of strawberries. I knew that by now she could buy oranges or even Nescafé from street vendors at a good price – but not strawberries. I bought them also because I remembered when we were together in New York for the first time, back in the eighties, and we went shopping. In a downtown Manhattan supermarket, we stood in front of a fruit counter and just stared. It was full of fruits we didn't know the names of – or if we did, like the man with the banana in Bucharest, we didn't know how they would taste. But this sight was not a miracle; we somehow expected it. What came as a real surprise was fresh strawberries, even though it was December and decorated Christmas trees were in the windows already. In Poland or Yugoslavia, you could see strawberries only in spring. We would buy them for children or when we were visiting a sick relative, so expensive were they. And here, all of a sudden – strawberries. At that moment, they represented all the difference between the world we lived in and this one, so strange and uncomfortably rich. It was not so much that you could see them in the middle of the winter, but because you could afford them. When I handed her the strawberries in Warsaw, Zofia said: 'How wonderful! I'll save them for my son.' The fact that she used the word 'save' told me everything: that almost ten years after we saw each other in New York, after the victory of Solidarity, and private initiatives in the economy, there are still no strawberries and perhaps there won't be for another ten years. She was closer to me then, that evening, in the apartment where she lives with her sick, elderly, mother (because there is nobody else to take care of her and to put your parent in a state-run institution would be more than cruelty, it would be a crime). Both of them took just one strawberry each, then put the rest in the refrigerator 'for Grzegorz.' This is how we tell our kids we love them, because food is love, if you don't have it, or if you have to wait in lines, get what you can, and then prepare a decent meal. Maybe this is why the chicken soup, cabbage stew, and mashed potatoes that evening tasted so good.

All this stays with me forever. When I come to New York and go shopping at Grace Balducci's Marketplace on Third Avenue and 71st Street, I think of Zofia, my mother, my friend Jasmina who loves Swiss chocolates, my daughter's desire for Brooklyn chewing gum, and my own hungry self, still confused by the thirty kinds of cheese displayed in front of me. In an article in *Literaturnaya Gazeta* May 1989 the Soviet poet Yevgenii Yevtushenko tells of a *kolkhoz* woman who fainted in an East Berlin shop, just because she saw twenty kinds of sausages. When she came back to her senses, she repeated in despair: 'Why, but why?' How well I understand her question – but knowing the answer doesn't really help.

3

Make-up and other Crucial Questions

When I close my eyes, I can still see her, resting on our kitchen sofa on Saturday afternoons. It's spring, she lies there in semi-darkness, and in the dim light coming through canvas shades on the window, I see her face, covered with fresh cucumber slices, like a strange white mask. She does it every Saturday – after twenty minutes she will get up, remove the rings, and wash her face with cold water. Then she will put on Dream Complex cream, the only cream available in the early 1950s. When I touch her skin, it will be fresh and tender under my fingers.

Before that, she would wash our hair, hers and mine. Her hair is very long and brown; she washes it with Camillaflor powdered shampoo. It's the only brand that exists and features brunette or blonde on the paper package. She rinses her hair in water with vinegar so as to leave it soft and silky. My thin, blond hair has to be rinsed with the juice of half a lemon – its smell follows me the whole afternoon. And while she lies there, with a cucumber mask – or a mask made of egg whites, or a yogurt mask if it is winter, or a camomile wrap to remove puffiness under the eyes after a bad night's sleep or quarreling with my father – I imagine that one day I will be doing the same, and I can't wait for this day to come.

Afterwards, I know she will manicure her nails and apply make-up. That is, she will put on one of the two shades of lipstick that she bought in a pharmacy and outline her eyes with a kohl pencil – these are all the cosmetics there are. She will be happy if she still has some powder left, packed in a small plastic bag, the one that her mother was able to provide for her. Finally, she will put a few drops of *eau de Cologne* behind her ears and on her ankles, a yellowish water made up by the pharmacist himself. In her blue satin evening dress that she sewed herself, she will go dancing with my father at the Army Officers' Club. Years and years later, I will remember every detail of how beautiful she was at that moment, a magician who created beauty out of nothing. She ignored reality, the fact that there was no choice, fighting it with the old beauty recipes that she had learned from her mother and grandmother. After all, one could always make a peeling mask out of cornflour, use olive oil for sun-tanning or as a dry hair treatment, or give a deep brown tone to the hair with strong black tea.

The young communist states at that time – and until they ceased to exist – had more important tasks than producing make-up, tasks like rebuilding countries devastated by war, like industrialization and electrification. Lenin's popular slogan was that electrification plus Soviet rule equals communism. In the five-year central plans made by men, of course there was no place for such trivia as cosmetics. Anyway, aesthetics were considered a superficial, 'bourgeois' invention. Besides, women were equal under the law, why would they need to please men by using all these beauty aids and tricks? However, even if hungry, women still wanted to be beautiful, and they didn't see a direct connection between beauty and state-proclaimed equality; or, rather, they did see one, because they had to work like men, proving that they were equal even physically. They worked on construction sites, on highways, in mines, in fields and in factories – the communist ideal was a robust woman who didn't look much different from a man. A nicely dressed woman was subject to suspicion, sometimes even investigation. Members of the Communist Party, for years after the war in Yugoslavia, had to ask for official permission if they wanted to marry a woman whose doubtful appearance would unmistakably indicate her 'bourgeois' origins.

But aesthetics turned out to be a complex question that couldn't be answered by a simple state decree. By abolishing one kind of so called 'bourgeois' aesthetic, not with a plan (except in China, where this kind of formal 'equality' was carried to the extreme in prescribed uniforms), but more as the natural result of ideology, the state created another aesthetic, a totalitarian one. Without a choice of cosmetics and clothes, with bad food and hard work and no spare time, it wasn't at all hard to create the special kind of uniformity that comes out of an equal distribution of poverty and the neglect of people's real needs. There was no chance for individualism – for women or men.

Once when I was in Warsaw, a friend told me about a spate of red-haired women: suddenly it seemed that half of the women in the city had red hair, a phenomenon that couldn't pass unnoticed. It might have been a fashion caprice. More likely, it had to do with the failure of the chemical industry to produce or deliver other kinds of dye. Imagine those women confronted by the fact that there is no other color in the store where they buy their dye and knowing that if there isn't any in one store, it's generally useless to search others. There is only the one shade of red. (I've seen it; it's a burgundy-red that gives hair a peculiarly artificial look, like a wig.) They have no choice – they either appear untidy, with bleached ends and unbleached roots sticking out, or they dye their hair whatever color they can find. so they dye it, hoping that other women won't come to this same conclusion. They don't exactly choose.

Standing in front of a drugstore on Václavské Náme'stí in Prague last winter, I felt as if I were perhaps thirteen years old and my mother had sent me to buy something for her – soap, perhaps, shampoo. The window of that drugstore was a time machine for me: instantly, I was transported into years of scarcity long past, years of the aesthetics of poverty. Even though I'm not an American, it seemed there was absolutely nothing to buy. In front of that shop window I understood just how ironic the advice in today's *Cosmopolitan* or any other women's magazine in the West is, advice about so-called 'natural' cosmetics, like olive and almond oil, lemon, egg, lavender, camomile, cucumbers, or yogurt. I still can recall my mother's yearning to buy a 'real' cream in a tiny glass jar with a golden cap and a fancy

French name, something she would have paid dearly for on the black market.

If Western women return to the old recipes, they do so by choice; it is one of many possibilities. Not so for Czech or Bulgarian or Polish women. I can see them arriving in Yugoslavia after days and nights in a train or car. They go to the market, put a plastic sheet on the street at the very edge of the market (afraid that the police might come any moment) and sell the things they've brought. Among them are professional black market vendors, women who make a fortune by buying foreign currency and then selling it back home for five or six times as much. But I also saw a young Polish woman, a student maybe, selling a yellow rubber Teddy bear, deodorant, and a green nylon blouse (the kind you can find only in a communist country or a second-hand vintage clothing store in Greenwich Village). I couldn't help thinking that she was selling her own things. But why would anybody in the world travel 1500 miles just to sell a plastic toy? And what if she sells it, what is it that she wants to buy with the money? Perhaps a hair dye that is not red . . . However, she is young, and there is hope that her life will be different. For my mother and women of her generation, it is already too late. If only they had had cosmetics, it might have changed their lives. On the other hand, it might not. But shouldn't they have had the right to find that out for themselves?

Once, when we used to play a childish adolescent guessing game, we would try to guess which of the women on the beach in Split were Polish and which Czech. It was easy to tell by their old-fashioned bathing suits, by their make-up, hairdo, and, yes – the color of their hair. Some-how, everything in their appearance was wrong. Today I realize that women in Poland like green and blue eye shadow about as much as they like artificial red hair – but they wear it. There is nothing else to wear. It is the same with the spike-heeled white boots that seemed to be so popular in Prague last winter. It is the same with pullovers, coats, shoes: everyone is wearing the same thing, not because they want to, but because there is nothing else to buy. This is how the state creates fashion – by a lack of pro-ducts and a lack of choice.

To avoid uniformity, you have to work very hard: you have to bribe a salesgirl, wait in line for some imported product, buy bluejeans on the black market and pay your whole month's salary for them; you have to hoard cloth and sew it, imitating the pictures in glamorous foreign maga-zines. What makes these enormous efforts touching is the way women wear it all, so you can tell they went to the trouble. Nothing is casual about them. They are over-dressed, they put on too much make-up, they match colors and textures badly, revealing their provincial attempt to imitate Western fashion. But where could they learn any-thing about a self-image, a style? In the party-controlled magazines for women, where they are instructed to be good workers and party members first, then mothers, house-wives, and sex objects next, – never themselves? To be yourself, to cultivate individualism, to perceive yourself as an individual in a mass society is dangerous. You might become living proof that the system is failing. Make-up and fashion are crucial because they are political. In Francine du

Plessix Gray's book *Soviet Women*, the women say that they dress up not for men, but to cheer themselves up in a grim everyday life or to prove their status to other women. In fact, they are doing it to show difference; there are not many other ways to differentiate oneself. Even the beginnings of consumerism in the 1960s didn't help much; there were still no choices, no variety. In fact, in spite of the new propaganda, real consumerism was impossible – except as an idea – because there was little to consume. Trying to be beautiful was always difficult; it involved an extra effort, devotion perhaps. But most women didn't have time or imagination enough even to try.

Living under such conditions and holding *Vogue* magazine in your hands is a very particular experience – it's almost like holding a pebble from Mars or a piece of a meteor that accidentally fell into your yard. 'I hate it,' says Agnes, an editor at a scientific journal in Budapest, pointing to *Vogue*. 'It makes me feel so miserable I could almost cry. Just look at this paper – glossy, shiny, like silk. You can't find anything like this around here. Once you've seen it, it immediately sets not only new standards, but a visible boundary. Sometimes I think that the real Iron Curtain is made of silky, shiny images of pretty women dressed in wonderful clothes, of pictures from women's magazines.' Fed up with advertising, a Western woman only browses through such magazines superficially, even with boredom. She has seen so much of it, has been bombarded by ads every single day of her life, on TV, in magazines, on huge billboards, at the movies. For us, the pictures in a magazine like *Vogue* were much more important: we studied their every detail with the interest of those who had no other source of information about the outside world. We tried to decode them, to read their message. And because we were inexperienced enough to read them literally, the message that we absorbed was that the other world was a paradise. Our reading was wrong and naive, nevertheless, it stayed in the back of our minds as a powerful force, an inner motivation, a dormant desire for change, an opportunity to awaken. The producers of these advertisements, Vance Packard's 'hidden persuaders,' should sleep peacefully because here, in communist countries, their dream is coming true: people still believe them, women especially. What do we care about the manipulation inherent in the fashion and cosmetic industries? To tell us they are making a profit by exploiting our needs is like warning a Bangladeshi about cholesterol. I guess that the average Western woman – if such a creature exists at all – still feels a slight mixture of envy, frustration, jealousy, and desire while watching this world of images. This is its aim anyway; this is how a consumer society works. But tomorrow she can at least go buy what she saw. Or she can dream about it, but in a way different from us, because the ideology of her country tells her that, one day, by hard work or by pure chance, she can be rich. Here, you can't. Here, the images make you hate the reality you live in, because not only can you not buy any of the things pictured (even if you had enough money, which you don't), but the paper itself, the quality of print, is unreachable. The images that cross the borders in magazines, movies, or videos are therefore more dangerous than any secret weapon, because they make one desire that 'otherness' badly enough to risk one's life by trying to escape. Many did.

In our house there was an old closet where my mother would stockpile cloth, yards and yards of anything she could get hold of – flannel, cotton, pique, silk, tweed, cashmere, wool, lace, elastic bands, even buttons. Sometimes she would let us play with this cache, but it was her 'boutique.' She would copy a blouse or a skirt from pattern sheets from *Svijet (World)*, the only magazine for women, and sew it on Grandma's Singer sewing machine. Every woman in my childhood knew how to sew, and my mother insisted that I learn too. By the age of five I knitted my first shawl and embroidered a duck with ducklings that she still keeps. Later on, she let me use a sewing machine under her supervision, and by the age of fifteen I was making my own dresses, not because it was a woman's duty, but because it was the only way to be dressed nicely. When, for the first time, she went to Italy to visit a relative there, she came back dressed in a white organdy blouse, a black pleated skirt, and high-heeled black patent leather shoes. She brought back a mohair pullover, a raincoat made of a thin, rustling plastic (it was called suskavac and everybody wanted to have one, since it was a sign of prestige), an evening dress made of tulle, covered with sequins that glittered in the night. What fascinated me most was her new pink silk nightgown and matching silk overjacket with lace lapels. So light, almost sheer, hanging down to her ankles, with two tiny straps that would leave her shoulders bare, it was the finest negligée I'd ever seen. I used to tell her that she ought to wear it to the theater, not to bed.

Mother's nightgown was for me the very essence of femininity. This was the first time, in 1959, that I'd seen that 'otherness' with my own eyes.

My mother brought something special for me too: three dozen sanitary napkins made of terrycloth and a belt. The napkins had buttonholes at each end to fasten them to the belt, so they wouldn't slip. She would hand-wash them, then hang them on a clothesline in the bathroom to dry overnight. More than thirty years later, in Sofia, my friend Katarina saw my package of tampons in her bathroom and asked if I could leave it for her. I am going on to Zagreb and she needs them when she has a performance in the theater. 'We don't have sanitary napkins and sometimes not even cotton batting. I have to hoard it when I find it, or borrow it,' she said. For a moment, I didn't know whether I should laugh or cry. I sprinkled Eastern Europe with tampons on my travels: I had already left one package of tampons and some napkins, ironically called 'New Freedom', in Warsaw (plus Bayer aspirin and antibiotics), another package in Prague (plus Anaïs perfume), and now here in Sofia . . . After all these years, communism has not been able to produce a simple sanitary napkin, a bare necessity for women. So much for its economy and its so-called emancipation, too.

Rumiana is a Bulgarian movie director and a member of the international organization of women in the film industry known as KIWI. In Bulgaria, KIWI operates like a kind of feminist organization, helping women in different ways, for example, by taking care of the children of women prisoners, helping out girls in reform schools and orphanages, and so on. Rumiana told me that she is 'in charge' of a reform school near Sofia. Every time she goes for a visit, girls there ask her to bring cotton batting. So she goes to a cotton factory, loads up her car, and then visits them. 'They are so grateful,' says Rumiana, 'even when it is something that they have a right to.' Today, when I think that my mother's silk nightdress doesn't necessarily have much to do with femininity, I still ask myself, what is the minimum you must have so you don't feel humiliated as a woman? It makes me understand a complaint I heard repeatedly from women in Warsaw, Budapest, Prague, Sofia, East Berlin: 'Look at us – we don't even look like women. There are no deodorants, perfumes, sometimes even no soap or tooth-paste. There is no fine underwear, no pantyhose, no nice lingerie. Worst of all, there are no sanitary napkins. What can one say except that it is humiliating?'

Walking the streets of Eastern European cities, one can easily see that the women there look tired and older than they really are. They are poorly dressed, overweight, and flabby. Only the very young are slim and beautiful, with the healthy look and grace that go with youth. For me, they are the most beautiful in the world because I know what is behind the serious, worried faces, the unattended hair, the unmanicured nails; behind a pale pink lipstick that doesn't exactly go with the color of their eyes, or hair, or dress; behind the bad teeth, the crumpled coats, the smell of their sweat in a streetcar. Their beauty should not be compared with the beauty that comes from the 'otherness.' Their image, fashion, and make-up should be judged by some dif-ferent criteria, with knowledge of the context, and, there-

31

fore, with appreciation. They deserve more respect than they get, simply because just being a woman – not to mention a beauty - is a constant battle against the way the whole system works. When in May last year an acquaintance of mine, a Frenchwoman, visited Romania (while there was still street fighting in Bucharest) she told me this about Romanian women: 'Oh they're so badly dressed, they don't have any style at all!'

Beauty is in the eye of the beholder.

4

I think of Ulrike this night in November

When I think of East Berlin, I can't help thinking of her first – twenty-seven-year-old Ulrike, tall, with long black hair, a strikingly white complexion, and two tiny, barely visible creases at the corners of her mouth that give her face a bitter expression. I think of the way she lifted her cup of coffee the first time I saw her, of my impression that she was not looking at me or at the street in front of us, but was turning her gaze inside at her own past, as if she were not really here, in Iowa City, in the States. She was there, in Berlin, it was obvious. But she didn't want to talk about it. Before Christianne, my West German friend introduced me to her, she warned me that Ulrike doesn't like to talk about 'that' – her escape from East Germany – and that I shouldn't try to press her. I didn't so much want to ask her how she escaped or how she had managed to end up here, in the middle of cornfields. I only wanted to ask her whether it is possible to forget, to change your life completely. Was there a new life for her? Ulrike hesitated. For a moment it seemed she was going to open up and talk, but then she slipped inward again, as if it were too hard to remember that life of three years ago, too hard to go back across the Berlin Wall, even if only in memory, to return to East Berlin, where she started.

It was a wintry morning in 1988. A sharp beam of sunlight was reflected back by the snow outside, cutting the semi-darkness of the café in two. On the other side of the sunbeam, as on the other side of experience, Ulrike drank her coffee in silence, distant and absentminded. Her remembrance, the unspeakable, carried her away to her room, the night before her departure to Hungary. She didn't say anything to her parents, but she looked at them that ordinary summer evening in 1984 as if she wanted to absorb every detail, every movement and word. This is how it happened, Christianne had told me: Ulrike and her friend took a train to Hungary and then to Romania. Everything went fine, because they didn't need a visa. Then they tried to go on to Yugoslavia – later, perhaps, to Austria or Italy – and they were caught at the Romanian-Yugoslav border. They both went to prison. Ulrike spent a year there. She didn't say much about that, she only told Christianne that she was so sick she thought she was dying – some rare lung disease. But there is one sentence that resonated in me for a long time, as I looked at Ulrike's nicely carved, ivory profile: 'She lost all her hair in the prison.' Her hair was long and healthy now, shining with the blue sheen that black hair sometimes has. It seemed so strong, so alive, so vital as she moved her head or ran her fingers through it. I thought of the horror conveyed by that sentence, the horror she must have gone through, and perhaps only then did I understand that there was no way for her to communicate it, no words.

When she was deported to East Germany, Ulrike was sent to prison again. Her parents and younger brother had to renounce her in order not to be tried as accomplices. Afraid for their position Ulrike didn't write to them, and she was sure she'd never see them again. Then West Germany 'bought' her out of prison, along with many other East German prisoners, gave her some money, and a place to stay. She started to live in West Berlin. And as soon as she saved some money, she went to India. 'Why India?' I asked Christianne in surprise. 'I don't know, perhaps because it is so far from Berlin, from East Germany, even from Europe.' In India she met an American professor. He teaches in Iowa City. They are going to get married. In the meantime – her hair grew back.

The next time I saw Ulrike was a week later. She called me and made an appointment to meet at a restaurant. 'I didn't want to be rude,' she said, 'but these things are difficult to talk about. I have a feeling I would have to "translate" my life into recognizable, familiar terms, otherwise nobody would understand me. But for me, there are no such terms. As for them there is nothing in their lives that can correspond to such an experience, except perhaps in the movies. So they look at me, and I recognize the blank screen in their eyes . . .'

In West Berlin, she worked in the Haus am Checkpoint Charlie, a museum at the crossing point to the East dedicated to the Wall erected on 13 August 1961, and to the memory of victims who died trying to cross it. 'What

haunted me there was that little home-made submarine
with which that young man – I clearly remember his picture in a bathing suit – crossed the Baltic to Denmark. He
was under water for five hours. I used to go and stare at it,
through a glass window, thinking of what would have happened if it had stopped, if guards had noticed him, of how
cold he must have felt, alone in the middle of the sea. It
was as if I myself was escaping from East Berlin every day,
again and again. I had to leave, I had to go somewhere.' As
Ulrike spoke hesitantly, slowly with her unmistakably German accent, her face changed. Through the window, she
looked at the main street, at the low brown-brick buildings,
at the department store, the movie theater . . . she didn't
belong in Iowa City. But where does she belong?

'Yes, I don't like to live here,' she said, more to herself
than me. Then she turned toward me: 'But if I have learned
anything from my life, it's that since I don't belong anywhere, only the movement matters. Traveling, being able to
travel, this is why I escaped, and what I enjoy more than
anything in the world is the fact that nobody is stopping
me.'

I met her in the fall of 1988. In the summer of 1989, I
stood on the observation platform at the corner of the Tier-
garten Park. I could see a good part of the Wall, covered
with graffiti on the West side, the wide, waste minefield
behind it, and then the Eastern Wall on the other side. On
the left side, the Brandenburg Gate and the Reichstag
Building, and on the far right, the Potsdamer Platz or what
was left of it after the bombing. Then I crossed to East
Berlin. From that side, the wall was completely white – a

long, white virginal strip of concrete, heavily guarded. I climbed the radio and TV tower near Alexander Platz, to the rotating platform, and together with a herd of Soviet tourists, I took a good look at the divided city. Even though one could hardly see the wall from 1000 feet above ground, one could easily tell where it was, especially in the evening, when the city looked like a large cake, cut into two parts, one darker and one lighter. Thinking that I'd seen enough, I went home.

Only three months later, on the night of November 9, people started to bring down the Wall. Like the rest of the world, I watched it on TV. As we toasted the event with champagne, I thought of Ulrike that night in the restaurant, of her lost hair. Is she relieved, finally cured, now that nobody would stop her from passing from one part of the city to the other, from one country to the other? Or perhaps it's too late for her to rejoice, because her life has been marked, and there is no way back, the end of the Wall will bring no forgetting? How must she feel now, when everything is over – if it is over?

A year later, in November 1990, I visited Berlin again. I went to the same place, near the Brandenburg Gate. But this time there were no walls, no platforms, no watchtowers – nothing, just a dark strip of new asphalt. A visitor who hadn't been here before would hardly be able to tell where the Wall had been – or even that it ever had been there – and I had to admit to myself that I hadn't imagined it like that. Then I took the 'Big City Tour East/West.' West Germans were quick to make money from the fact that there was no more Wall. The bus, a double-decker leaving from

Ku'damm, was full of Dutch students and a couple of older people, foreign tourists it seemed. As we were approaching East Berlin from Kreuzberg, even without a visible border, I felt a strange uneasiness. I guess everyone does, or at least everyone who even once has stood in a long line at the Friedrichstrasse stop, the only possible exit – or entrance, if you prefer – to East Berlin from the subway. After a long walk along underground corridors, you come to a great hall traversed by a wall of passport control kiosks. Their narrow doors allowed only one person at a time to go through. There you waited – you could never tell for how long – until finally your turn between the two opaque barriers arrived. Then a policeman took a long look at your passport, at you, at your passport again, while minutes passed like hours. It was this look that always irritated me, the familiar look of a suspicious officer at every communist border. You know you are not guilty, but you also know that it doesn't matter, because he thinks you are and because you are entering a world where there are no innocents, where everyone is guilty until proven innocent. The very fact that you are traveling from one country to another is bad enough. Once in the other part of the hall, entering East Berlin, the world would look different – and you would look at it differently.

From the border the double-decker bus entered Unter den Linden Avenue and stopped at Marx-Engels Forum, in front of the now deserted party building. An old couple asked me to take their picture under the street sign. They were West Germans. I didn't need to ask them why they wanted this picture taken, it was so obvious. But as if they could read my mind, the woman said, 'It'll be gone soon!' Listening to her, it suddenly seemed as if I were not in a real city, but on a huge stage where they were about to change the scenery, once more after forty years. As I held the camera, I realized that soon this very photo I had just taken might well be the only proof, besides our weak memories, that streets named for Karl Liebknecht or Clara Zetkin or Marx or Engels existed once, not so long ago. It will be as if someone with a sponge is walking slowly from street to street, from square to square, wiping the names out, removing the material evidence of the life before.

Next our bus stopped at Republic Square, in front of the Reichstag Building. For the first time right in front of me I see the building I had known only from movies about fascists: it is huge, dark, and frightening. On the left side of the Reichstag, where the Wall had stood, there are a few white crosses. People put them there after the Wall was torn down, in memory of those who were killed on that very spot – the last one in the spring of 1989. With the Reichstag in the background, the crosses look small, pathetic, fragile. Even the flowers are withered. 'We are thinking of moving the new government of the united Germany back here again,' says a guide, a neat young man in a white shirt. He says it casually, letting his flat, cool voice fall over the empty square, the gray River Spree, the void in place of the Wall.

Again this November, I am thinking of Ulrike. What would she feel right now, in front of these white crosses, the absent Wall, the Reichstag Building with the German flag fluttering happily on the wind? Would she feel robbed of something – of her past, perhaps? Two years in prison, her illness, her hair falling out, her fear, persistence, determination – all for nothing, all in vain? So much suffering left to the vagaries of memory, reduced to a bus tour to the East – while it still exists – to the indifferent voice of a young guide. It is not that I mind the demolition of the Wall – I am delighted about that – but the way it was done, the obvious haste with which this tumor was removed not only from the face of the city, but from the memory of the people, too, acting as if it is really possible to unite instantly, to become one Berlin, one nation, as if the past, the division of that nation, doesn't count at all anymore and should be instantly forgotten. As we were passing by an ugly, military-looking building, the guide said, 'This used to be the Aircraft Ministry, and a very famous person, Mr Hermann Goering, used to work here.' This is it, I thought, the erasing of memory begins right here, right on this spot near the Potsdamer Platz, right when Goering is reduced to a *very famous person*, and the Wall to tiny bits of painted concrete selling for 5 Deutschmarks, when the whole history of this nation is reduced to souvenirs and fame. What I feared is already here: incoherent bits and pieces of the past that don't make any sense anymore. That, in fact, are not important. But the sooner we forget it, the more we'll have to fear.

Instead of being turned into a symbol, a monument, Checkpoint Charlie – that mythical door to a better, imagined, utopian life, where people were killed as they tried to pass the barrier, where they waited to get across just to glimpse the other kind of life, something that they called freedom and were prepared to pay for with their lives to taste it – is a flea market now. The lumber room of history, where people, many of them not even Germans, but Turks and Gypsies, sell old German Army uniforms, badges with hammers and sickles, and nicely packed fake pieces of the Wall – a new supermarket selling leftovers of the dead regime. After I saw this, I thought I wouldn't be surprised if tomorrow they made a flea market out of Buchenwald, too, selling the hair of prisoners as 'souvenirs' (with a certificate of authenticity, of course).

But Ulrike's museum is still there, perhaps the only decent authentic place that keeps alive the memory of the horror. And thank goodness, because this is something that we badly need to remember, not to forget, if we are to preserve the past, however horrible it was. This museum is small, modest. Somehow, it is not impressive, it doesn't fill the whole word 'museum.' Perhaps the greatness of it lies in its very plainness, in the extraordinarily banal, primitive, simple improvised devices people used to escape and the hope they invested in them – the same objects Ulrike had tried to describe to me four years ago and that had sounded so incredible there, in that restaurant in Iowa City. They were not for real there. Here, at the old border between two worlds, is the only place in the world where the inventions of captive minds – as well as Ulrike's words – can be real, can have real meaning. The mini-submarine with the 28-inch hollow plastic body, a small motor, and a propeller (the note says that the young man who defected made a fortune by producing them later on); the first home-built escape aircraft, with a Trabant car motor; the hot-air balloon; the chair-lift and rope with which a whole family escaped one dark night in 1965 from the roof of the Haus der Ministerien; the tiny little wooden pushcart with which sand was lifted from a 500-foot-long tunnel. . . It is in the face of this genuine desire for freedom, and the fragility of the means to reach it, that I felt the Wall shouldn't be wiped off the face of the earth. It should stay there – not whole, but big chunks of it, in all its absurdity, as a vivid, solid monument of the past – a monument of division, suffering, terror, injustice – in the name of people who were killed and of the generations that lived in its shadow for almost thirty years. The 'very famous person, Mr Goering' and this museum are two poles of German history, with people's lives stretched in between.

On the second floor of the museum there is a small room, seven feet by seven feet, with several TV screens showing a picture of the Wall – and a little window that had overlooked the real Wall. But looking through this window in November 1990, one could see nothing, just empty space. They say that sections of the wall are recycled for sand at a factory in Bernau. As I stand here, in this small cabin in the museum, looking at the void, I imagine a fine, white sand from Bernau falling down on the land, covering the past with a white silence, as with snow. In Iowa City, Ulrike leaves her house, thinking that snow is falling. But when she looks more closely, she can see that it is a fine, white sand, almost a dust – dust of the Berlin Wall.

5

On Doing Laundry

As if I am looking at them today, I remember my grandmother's hands: swollen and cracked, with little sores from washing dishes, laundry, floors, windows, kids, from constantly dipping them in hot and cold water. She was our servant – and our washing machine. Every Saturday she would perform a laundry washing ritual, a very long and elaborate procedure. On Friday night she would soak everything in the big metal washing tub. On Saturday morning, she would first scrub it, bent over the rim, using a wooden washboard. Then she would put it on the stove and boil it for a while – that is, the white cotton and linen sheets, towels and underwear – stirring them with a long wooden stick from time to time.

After the laundry had been rinsed three times and whitened, she would starch it. First, she would cook starch in a small pot over the stove, then drop it into the water, so that it looked almost like milk. 'This will make it look nicer, and it helps wash the dirt off later on,' she would explain to me, even though I was only seven years old (but a girl of seven should know these things). On Sunday evenings she would iron the whole pile of perfectly white clothes, stiffened with starch. She did it with an iron heated by coal that I still must have somewhere around the house; she would show me how many times you should fold a sheet and how to iron a man's shirt. This was in 1956, and I still can recall the smell of 'Women's Thanks' (Ženska hvala) detergent in the bathroom, the steamy kitchen windows, how she would test the iron with a wet index finger, and the way she would slowly rub her hands with glycerine afterwards.

After all was done, she would sit by the kitchen table, tired, looking at her hands, then at me, as if she were thinking: Poor child, this is waiting for you, too. Once, much later, I asked her why she didn't use rubber gloves. She looked at me in bewilderment. 'You silly girl! Do you really think there were gloves to buy?' But she was very proud of her laundry. She would hang it outdoors, in the backyard and on the balcony, and it was the whitest in the yard – actually, it was slightly bluish from the tiny balls of bluing she would drop in to dissolve in the third rinse. When we took it down from the line together, she would hand it to me, and I would dip my face deep into it, smelling the north wind, the sun, the very cleanness itself.

Because of Grandma, my mother escaped her destiny. When Grandma became sick with rheumatism and couldn't wash any more, Mother would take all the clothes to the state-run laundry, where a serious looking woman dressed in a white coat, as in the hospital, would hand her a yellow piece of paper with a number on it. In a couple of days mother would bring the ticket back and collect the wash (they didn't collect or deliver clean laundry). The laundry was clean, or so I thought, but it didn't smell as nice, and it had ugly, black numbers stamped in the corner of every sheet and towel, which made you feel as if it wasn't yours any more. Grandma would put it back in the closet, with disgust showing on her face. 'Oh, they've ruined everything!' she would say. 'Look how gray my fine batiste sheets are. They spoil everything, everything, those thieves!' She sighed, angry at 'them.' It was unclear who, in her imagination, 'they' were – the state, the government, the party, or only the laundry people? And nobody knew why she called the workers at the laundry thieves, since they had never stolen anything from us. I guess she generally felt robbed because her big house in Rijeka had been nationalized after the war. But her beautiful hand-embroidered sheets she never sent to the laundry. Instead, she decided she would rather not use them any more. 'They'll be yours,' she told me. They are now. But I'm not using them, either, afraid that after washing them in the machine they won't be as white as she would like them to be.

A couple of years ago, *Time* magazine published a cover story on women in the USSR. Describing the hardships of women's lives there, a reporter wrote: 'Clothing is generally laundered in tubs, then hung out to dry.' Thinking of my grandma, I was puzzled by this sentence for quite some time. I just couldn't get its meaning right. Am I to be mad at the explanatory and, as it seemed to me, patronizing tone of this sentence? Doesn't it sound a bit too much like an anthropologist's writing about the strange customs of a

45

jungle people? Why is the reporter, a woman, so surprised that laundry is done in big plastic or metal tubs? Is that so unusual, so strange, considering that it was done in just the same way in the States fifty years ago?

In a way, the reporter was an anthropologist, going back in time, as if describing long-forgotten customs in some remote part of the world: 'Disposable diapers are unknown. Floors are scrubbed with brooms wrapped in damp cloths . . .' She could have gone on writing in the same vein, describing women kneeling on the floor, their red swollen hands without gloves, the pain in their backs, their tired, pale faces. It would be the truth. There are no appliances, no services, and no agencies in the USSR where they could hire a person to help, and women are, regardless of class, sentenced to long hours of repulsive housework everyday. Because this is a serious class question, there are no pre-revolutionary *slugi* (servants) in communist countries to serve the bourgeois class – there is no bourgeois class. We are all equal. Why should one member of the working class clean the apartment of another? Wouldn't it be 'exploitation of man by man'? But then, what is one supposed to call hand-washing of laundry, scrubbing floors, or ironing? The answer is: just women's work. It is not that the state hated women and, therefore, didn't produce machines that would make their lives easier, but rather that there were so many other problems to solve, things to produce. The 'woman question' (if any!) was going to be solved one day, that's for certain. Women just had to be patient, had to understand the vision of the great revolutionary plan, a vision in which their needs – what with Ideology,

Politics, and Economics – were nowhere near the top. It was almost self-evident that, once these great, basic problems were solved, all *their* problems, even floor scrubbing, would be solved too. I can almost imagine that great day, when every woman in the Soviet Union (and in every other communist country) would wake up to find in their bathrooms not only washing machines, but dryers, hairdryers, electric toothbrushes and shavers, vibrating shower heads, and so on. In the kitchen, they would find dishwashers, toasters, mixers, microwave ovens, little espresso coffee machines, electric can openers, wine coolers, deep freezes, Cuisinarts. ... I mustn't forget the latest model vacuum cleaner that, perhaps, washes windows too – and an automatic floor polisher, if there is such a thing. But the special present from the state, once it could, finally, in the year 2084, address the 'minor' problems at the bottom of the list, would be a huge, brand-new refrigerator, stuffed with food!

On the other hand, perhaps the American reporter was simply sympathetic, feeling sorry for the Soviet women, who had to do laundry without household appliances. It is truth that even if there were money, the family would be more likely to buy a color TV than a washing machine. Finally, I concluded that what disturbed me was the fact that the reporter had to explain to her American readers that there is still a part of the world where clothes are washed in tubs, by hand, and hung outdoors. For them, it is not normal, it is not natural – it is outrageous. In that sentence came together a genuine surprise that the communist state, even after seventy years of existence, was not able to produce enough cheap washing machines and a profound ignorance of the fact that three-quarters of the world still washes clothes by hand. Even more, one sentence in *Time* magazine reveals all the cultural, social, and economic differences between American and Soviet society, between the capitalist and the communist systems, turning them, for the reader, into two different cultures.

Even though I'm not a *Time* reporter, I was taken aback when I first entered my friend Ewa's bathroom in Warsaw in the winter of 1989. There it was, the very first washing machine we ever had. I was nine years old when my father brought it home proudly, unwrapped it, and installed it. It was a very simple machine that didn't have a spin cycle. To wring out the clothes, you had to run them through two rollers, turning a crank. Grandma was suspicious, my mother delighted. And here, in Ewa's small bathroom, it was again, an antique washing machine, belonging more in a museum than in a modern household. Ewa shrugs her shoulders. This is it; she simply can't buy a new one. She is a single mother – and besides, this one still works. This is precisely the same way I think: my primitive automatic machine is almost twenty years old, it bumps and squeaks, but it still works. It has sentimental value also: it was given to me by my mother-in-law when my daughter was born. At that time it was a great investment, but she – a former teacher and a very emancipated woman – thought a washing machine took priority over any other home appliance, and I'm still grateful to her for that. Having a washing machine was a luxury for another reason. In the cellar of our building there was a washing room with a huge concrete washing basin and three new washing machines. At the beginning, everyone washed their clothes downstairs. There was a schedule hung on the door and each family took its turn once a week. The machines didn't work for long. To put it mildly, people didn't take very good care of them. After all, these machines didn't belong to anyone in person, so no one felt responsible for repairing, or even cleaning them. The first machine broke after about a year, then the second one, then the third. In the washing room, people started to store broken chairs, children's bicycles, beach umbrellas, charcoal for barbecues, skis, mattresses. . . . The basins were filled with supplies for winter: bags of potatoes, green and red peppers, and wooden barrels of sauerkraut.

We'd lost our common laundry room precisely because it was common. But by that time the standard of living in the country was high enough so, instead of forty people using three common machines, everyone could buy an imported washing machine for themselves, however unnecessary and irrational this really was. Even our own country started to produce them, except that they all were very expensive. This, strangely enough, became a reason to buy one, to prove that you were earning enough, that your social status was high enough, so you could afford household appliances. Social differentiation, starting with cars and TV sets, continued in bathrooms and kitchens. A washing machine became an item of prestige, and it was good for women, even if it wasn't really meant to ease their lives in the first place.

On the ground floor in the building where we lived, there was a family that recently moved in from the country.

They owned a little truck and became rather affluent by buying vegetables in remote villages and selling them at the farmer's market for double or triple the price. It was not quite legal, but they managed somehow – perhaps in the same way as everybody else, by bribing the authorities. The story in our building went that when they moved in, the wife built a fire on the bathroom floor, because she had never seen a bathroom before. My mother told me that after the war, people used to keep pigs and hens in their bathrooms or put soil in the bathtub to grow carrots or lettuce ... Soon enough, this family bought a washing machine. The wife still washed laundry by hand, but it was one of the signs of how rich they were. They kept the washing machine in the kitchen, covered with an embroidered tablecloth, showing it proudly to every visitor. It took some time for the wife to realize that this was something she could and should use.

'Even if I had a modern washing machine, I couldn't use it very often, simply because there's no detergent to buy,' says Blaga. She recently had a baby out of wedlock and lives in Sofia, in an apartment with her mother. 'When I find it, I buy two or three big boxes. You can never be sure when it will appear again. I am doomed to wash by hand. I am saving for a washing machine, though, and I hope that by the time I buy it, there will be more detergent. The point is I won't need it so badly as I do now.' She washes her laundry with the same washboard that my grandma used, only hers is made of yellow plastic. When I tell her that, she smiles. 'Well, you can't say that there is no progress in communism!' But actually, she is disappointed, very well aware that because of the political and economic problems of her country, her problems seem small even to the new government. She lives on the third floor, and because she doesn't have a balcony to hang her clothes on, she has a device that I've seen so many times, on so many windows: two metal tubes fixed either under the window or on the window frame itself, with rows of lines in between. The laundry hangs above the sidewalk, water dripping on the heads of passers-by. 'People don't mind it, everyone does it. After all, we have to hang laundry somewhere, don't we?' says Blaga matter-of-factly.

Well, perhaps the answer is a dryer, but Blaga hasn't heard about such a machine. Perhaps the architect of her building has. I imagine him as an energetic, modern young man who, ten years ago, imagined that dryers would soon become a normal thing in socialist Bulgaria – so why build balconies? Apparently he was too optimistic. More likely, he didn't even know that laundry has to be hung somewhere, because he is a man. To tell the truth, the huge building that he designed has some tiny, tiny balconies, every fifth apartment or so. And the way they look makes me think that *they are dryers*: I have rarely noticed anything else on them, besides laundry. This, I think, looking down from Blaga's window in Sofia, is what makes our cities so specific, so unique - balcony dryers. I wonder why my colleague from *Time* didn't write about that too, about where and how people hang clothes here, about the whole world of ropes up in the air. Perhaps you don't notice it at first, in the center of the city and on the main streets. But as soon as you enter the side streets, hanging clothes flutter like flags of another state, announcing that you are entering a different, female territory. Clothes dangle on the wind under the windows, on balconies and terraces, in backyards, in narrow streets stretched between houses, even high up on skyscrapers. Socks, pants, shirts, diapers, dresses, aprons, handkerchiefs, slips – they make a foreign city all of a sudden look intimate, friendly, familiar to me.

By looking at the clotheslines I can tell who is a good housekeeper, whose laundry is white enough and properly hung, how big the family is, who lives alone. My grandma taught me all that, and how to hang a man's shirts, trousers, or pullovers. She taught me that you could tell a lot, even the character of people, just by looking at clothes. Laundry was like an open book to her. 'That woman over there, she is playing the lady. Look how many nylons she's hung out!' Or, 'The one from the first floor, she must be stingy, she doesn't use bleach at all.' We would walk down the street, and she would suddenly point something out to me: 'There, behind that window, lives a young woman with a baby, you can tell it by the dozen diapers, the nightgown and bra. It's a baby boy, his tiny pullover and cap are blue. A man lives there, too, perhaps the husband, judging from the shirts and socks. But they don't live alone, there is an older person. See the black woolen dress, still dripping? Perhaps she is a widow.' She taught me to observe, to look carefully at things around me, however small and unimportant they might appear to be, so I could learn more about people. In that way, she told me, 'In time everything will be revealed to you.' Maybe because of her, today I am not tempted to buy a dryer. I think I will always hang my clothes outside for the sheer poetry of it, so I can take them down from the line, and feel the smell of wind. In fact, I think this is the only way to smell the wind – with your face buried deep in freshly dried sheets.

Much, much later, in the middle of the 1970s when we already had a modern washing machine, Grandma wasn't pleased again. She insisted that no laundry could be washed properly without being boiled, and the machine could only heat water to 95 degrees Celsius, five degrees below the boiling point. But it was only when I first washed my clothes in the States, in 1983, in an American washing machine, that I became aware how differences in tradition influence both the industry and my own attitude towards doing laundry. American machines are simpler, they have only a few programs, they don't heat water even to 95 degrees Celcius and – what really astonished me – their cycle is at least three times shorter than any of the European machines. Even though my grandmother had already passed away, I just could not help remembering her, because, strangely enough, I felt as if my clothes were not properly washed at all.

But one thing is certain. In spite of progress, democracy, consumerism, and all the wonders that the new government is promising, I will keep my washtub. Five or six times a year in Zagreb there is no water. Five or six times a year there is a 'reduction,' a cut in electrical power. Some four or five years ago, for example, we spent every second afternoon or evening in total darkness, because of the shortage of power. That is, if you don't count candles. So, the sale of candles skyrocketed, but the sale of tubs did, too. As it was, we could well use coal-heated irons. It happened under the communist government, but just the other day the new democratic government announced power shortages again. This is why a woman needs to have a tub (and candle) ready even today, because she doesn't know what is waiting for her – she has no way of knowing what democracy will look like in an Eastern European country.

6

A Doll that Grew Old

It was in West Berlin. One early Sunday morning in February, while a thin crust of ice crunched under my feet, wrapped in a milky mist, Berlin was still sleeping. I went to visit the fleamarket near Potsdamer Platz. The Poles come early. They occupy the market, selling cheap tents, sleeping bags, thermos bottles, fur hats, old-fashioned fans, and record players. Behind the wire fence that surrounds the market – they called it the 'Polish market' – I saw a doll. She was lying there, on a piece of dirty cloth spread over the frozen ground, among silver spoons, a kerosene lamp, two candlesticks, and a set of screwdrivers. But even amid the clutter, I immediately spotted her beautiful little head made of lacquered papier-mâché, her delicate face with rosy cheeks as if she had a fever, her carefully painted blue eyes and tiny red mouth. She was dressed in a blue-flowered dress, crumpled and shabby, that hid her anemic body sewn of cloth. But I recognized her because she looked like the doll I had a long time ago, the first doll my mother bought for me, God knows where, because in the early fifties there were no dolls to buy in our wrecked country. A strong, blond woman with a heavy accent asked 'Zehn Mark,' but I gave her five. When I lifted her from the ground, the doll smelled of dust, of mothballs, of the bottom of an old drawer.

After more than thirty years, it felt so good to hold her in my arms again, as I did in 1956 when I traveled on a boat to Italy, across the Adriatic Sea. I was sitting on the upper deck in a chaise longue eating a pear, and although Grandma was sitting next to me, I was a bit afraid. The day was clear and I could hardly tell the blue of the sea from the blue of the sky on the horizon. It felt as if I were floating in a transparent crystal sphere over the sea, swinging slightly on the wind, as if I were unreal. Only the taste of pear in my mouth and the doll's body pressed tightly against my chest gave me the sense of reality, of security that I needed. I took her on this trip even though my mother said I shouldn't because I was too big now. But I insisted. 'At least I'll make her a new dress,' she said. It was a pink organdy dress, and I thought she was the most beautiful doll in my neighborhood – that is, in the whole world.

My aunt and uncle were waiting for us in Ancona. On our way to Naples, they had to stop the Fiat every half hour or so because I was not used to riding in a car – only a bus, (there were almost no private cars in Yugoslavia at that time), and I was vomiting all the way. I remember driving in a car on just one occasion before. It was when my father, an army officer, took us with him on a business trip. We had to travel very slowly because the road was gravel. But even so, I was so sick that I was convinced that this was it and thought how sad it was to die in this awful, stinking, shaking black monster . . . As we drove through the lovely landscape of southern Italy, I suppose my grandma didn't see much of it. 'Poor child,' she sighed, holding my forehead, probably pitying me because I was sick, but also because I was not fortunate enough to have been born in a country where cars were normal. But we did reach Naples alive, and by the second day I received enough presents from relatives excited that someone from a communist country had come to visit them that I soon forgot this unpleasant episode.

I don't recall this, but my aunt, who will be eighty this year, swears even now that this is what happened. We went shopping, because she wanted to buy some clothes for me. 'She looks like a beggar in this coat!' she told my grandma, with a tone of reproach. So we went to Standa, the big department store. As we were passing the toy department, I spotted a big doll seated in a doll's chair. I stopped in front of her, looking as if I had seen a ghost. My aunt tried to pull me away, but I just stood there, staring at her, without word or movement. I had never seen such a creature, not in my wildest dreams. She was the biggest and most extraordinary doll, different from mine or any of the other dolls I'd seen – even though I had not seen all that many. My aunt then let me look at her for a moment and went to buy something. But when she came back, I was still standing there, in front of the doll, in the same position, as if hypnotized by her eyes, which looked so real. My aunt couldn't do anything else but buy her.

Back home, she unwrapped her, but I barely dared to touch her, looking at her with the same fascination, not understanding that she was mine now. In the Yugoslavian shops there were almost no toys to buy and, until that moment, I was convinced that my papier-mâché and cloth doll, in her new dress, was unique. 'Real hair!' exclaimed my grandma, when she saw the new doll, almost as excited as I was. Tall to my waist, the new doll had black hair and blue eyes that, to my utter amazement, shut when I put her down on the bed. But this was not the only miracle she performed. When I turned her upside down, she would cry 'Mama!' In her dress of white silk batiste embroidered with little red roses, with a white petticoat and underpants, real shoes that one could unbutton, and white stockings, she was so different from anything I'd ever seen that she was a sheer wonder. I laid her down on the bed, touching her face, lips, ears, her fingers with red nails, as if I were not sure whether she was alive. Then I turned back to look at my old doll, the one that had traveled with me and – as I believed – saved my life during the torture in the car. All of a sudden I was furious. She looked small and incredibly worn out. How terribly naive I was, to think that this dirty little pile of papier-mâché and cloth was the most beautiful doll in the world. I felt let down, as if she had betrayed me somehow, as if she personally was responsible for being so ugly and that the best thing would be to forget her, leave her behind.

Back home in Zadar, I suddenly became the most popular girl in the neighborhood. Girls from my class would come to my home after school just to see the doll. Or rather, not just a doll but an icon, a message from another world, a fragment of one reality that pierced into the other like a shard of broken glass, making us suffer in some strange way, longing for the indefinite 'other.' We didn't know what it was, but it certainly looked beautiful. The doll would sit in my rocking chair in the middle of the room. I would let girls look at her, and maybe touch her if they would give me candy – but not play with her. My best friend Kristina, was the only one who could hold her and even turn her upside down, so she could hear that magic voice. But it was my privilege to play with her, and for the first time in my life I felt that I owned something important – something that in the eyes of the others made me different. The problem was that I didn't really play with her. She was too perfect, too . . . special perhaps? If I played with her, I might spoil her dress, mess her hair, or even break one of her legs or arms. Besides, I was not used to the stiffness of her body. I couldn't squeeze her, take her into bed with me, bathe her, or even hit somebody over the head with her.

I missed my old doll, and I felt guilty because I had left her for the new one, which was beautiful but too distant. I discovered she was not a doll to play with. So, I had to invent other dolls to play with. Because they didn't have dolls, girls in my apartment building were used to inventing them. An older girl – her name was Snježana and she was thirteen – would draw a paper doll for each of us about eight inches tall. She would cut it out, then glue it to a piece of cardboard with Karbon glue that smelled of almonds, then cut it out again. These dolls didn't last long, but I loved them because I could have three or four, in fact as many as I could persuade Snježana to draw for me, or as I could copy myself. And I could choose the color of their hair and eyes, their hairdo, or their bathing suit. Snježana drew all of them in one-piece bathing suits; it must have been the fashion and a norm of the time. None of us thought to ask her to draw a nude one.

But the best thing was that I could make a lot of paper dresses for these dolls. From Italy I had brought a big set of watercolors, and in their brightly colored clothes copied from my mother's fashion magazine *Svijet*, my dolls looked as if they had just stepped down from its pages. We dressed them in all the clothes we wanted for ourselves but didn't have: coats, pants, pleated skirts, pullovers (wool was difficult to find), all in the latest style or in long skirts – and this I liked the best – evening dresses with flounces or lace and deep necklines. We even made little accessories like handbags that matched their tiny paper shoes, gloves, necklaces, hats with veils that would give them – as we imagined – a mysterious, womanly look. These dolls had another advantage over the others: they grew as we did. At the beginning they didn't have breasts, but later on Snježana would draw them – rather shyly – and we didn't dress them in polka dots or floral dresses with ribbons for little girls anymore, but in the long golden lamé dresses that we saw only in rare American movies.

On Sunday mornings we would sit in the lobby of our building and play. We would show off their new dresses, change them, pretending that our dolls were going to work or shopping or to the movies. My dolls, I remember, were either hairdressers or vendors in a newspaper kiosk. Our little world was populated by dolls who, in our hands, behaved like real women. Except that there were no boys around. We played out the utopian situation where men didn't exist or were not important. Our dolls were self-sufficient just as we girls of six, seven, eight were self-sufficient at that time. On the other hand, we wanted them to be pretty, even if nobody told us why a doll (a girl, a woman) had to be pretty. We just knew it had to be so. We painted their little lips and nails bright red, and dressed them in tight sexy dresses, even if we didn't know what it was all about.

They were a reflection of our reality. We realized that one day we would grow up and we wished, when that day came, to be beautiful. Secretly, Kristina and I painted our lips and eyelids with my watercolors, wrapped shawls around our thin bodies, imagining we were the Tahitians in *Mutiny on the Bounty*. Even so early, even in postwar Yugoslavia, the standard of beauty was already set for us and – hard as it is to believe – it came from Hollywood. There was this magazine, *Filmski svijet* (*Film World*), and soon we were copying the faces and curves of Rita Hayworth, Ava Gardner, or Brigitte Bardot for our dolls. These dolls we hid from the eyes of our parents – we knew that there was some reason to hide them: we were not allowed to read *Filmski svijet*. Sometimes I think that at that early age I learned everything about my sex from these thin paper dolls – or, at least, the basic knowledge that for a woman it is more important to have a consciousness of her body than of her self. Later on, it took me – and our whole generation of women – years and years of hard work to unglue ourselves from those paper idols, to break through into another dimension, away from the dolls of our childhood, to which we were constantly reduced.

I was growing up and, in the meantime, the big doll sat in my bedroom like the queen of all the dolls in the world, superb in her impeccable dress, in her untouchable beauty, out of our world. Only for birthday parties would I take her out, just to keep my exclusive reputation. Then gradually dolls disappeared from our world, populating drawers, closets, cellars, and cardboard boxes. In no time I was married and had a little girl myself. Very anxious to avoid the pattern of womanhood I lived out when I was growing up, I educated my daughter 'neutrally,' and she played with toy animals instead, trains, cars, Teddy bears. Once when we visited my mother, we found the queen-doll in the closet. I took her out. There she was, smaller than I remembered her, in the dress that had once been white but was now a dirty yellow, her shoes lost; one of her blue eyes wouldn't open anymore. I looked closer at her face and for the first time since I saw her in the Italian department store, I realized that she too had changed, that she too had grown older. On her face, I could see time. More than anything else, even the birth of my own child, it made me understand that my childhood was over. 'You know,' I told my daughter Rujana, 'I never, ever played with her. Not for one day. She wasn't a toy to me.' Rujana, who was about five years old at the time, gave me a strange look. I could tell she didn't understand what I meant – to have a doll and not to play with her, what a funny idea. 'But why?' she asked, starting to undress the old doll without any sense of how special she was to me. She was too little to understand what I myself had understood only recently, that the doll was too precious, too important, too distant from me to be a toy and to participate in my childish games. She represented that other world I had glimpsed just for a moment, the ideal. 'I preferred paper dolls,' I said, trying to excuse myself. 'What are paper dolls, Mama?' Rujana asked. I drew one for her, but she wasn't impressed.

Until the other day, I believed that, with my help, my daughter would escape the traps and clichés of the Hollywood idea of beauty, so she could be more free in growing up to be a woman. But only a couple of days before her twenty-second birthday, when I asked her what she wanted for a present, she told me she wanted a Barbie doll. I wasn't surprised – I was simply shocked. It seemed that my feminist education was in fact a house of cards, if my daughter, whom I consider to be smart and emancipated, wanted a Barbie, the symbol of bimbodom. Then she told me this story. When she was about ten years old, she had stolen a Barbie doll from her friend and hid her in her room, under the bed. For two weeks she played with her every afternoon. She even packed her, addressed the package to herself, and sent it through the post office, pretending that this was a present from her aunt in America. At the time she was staying with her grandparents on the coast, and they knew perfectly well she didn't have any aunt in the States. Moreover, they knew this was the stolen doll. But even when they told me about it, I didn't realize how badly she wanted a Barbie and, what is worse, I had forgotten the whole episode. 'Even after that, you didn't buy me one. Didn't you remember your own dolls, your own longing?' my daughter asked, making me feel guilty twelve years later. I didn't answer, I felt too ashamed. Of course I remember them, and I know I did her wrong. I just couldn't buy her a Barbie, it would be against all my principles, against my ideology. I didn't have dolls because of the poverty that came out of a communist ideology – but she didn't have them because of an ideology, too. In denying her a Barbie, was I more right than the communists? Perhaps I was, but it hurt my daughter. She is emancipated now, but she had to pay a toll to an ideology as well; I realize that only now. It took me time to see that any kind of ideology could reduce us to poverty and emotional suffering.

I knew it was too late, but for her twenty-second birthday I bought her a Barbie. I think dolls are important – the dolls that we had, that we didn't have, that we longed for, that we betrayed and left behind. At some point in our lives we have to return to them, to rethink their meaning, or go back to them for a brief moment, just to restore the feeling of intimacy and security they embodied for us. I think it was the security I wanted to feel once more, as I stood there, at the fleamarket in West Berlin, on a foggy Sunday morning, holding a doll again after so long. The world I knew was falling apart. We were all traveling toward the unknown – wanted, but unknown, maybe too beautiful to live in, like my doll, which was too beautiful to play with. I wonder – this feeling of insecurity, of vulnerability, of having the carpet pulled out from under one's feet, this fear in the face of the desired but unknown – is this the price that we have to pay to reach toward a future that is constantly escaping us?

7

Forward to the Past

'No!' she said resolutely. 'I'm not going to use it!' I stood for a moment in front of the shelf in the drugstore, holding a package of cheap toilet paper in my hand. 'My God,' I almost screamed, 'how I would like to slap your face! Where do you think you live? I grew up without all those fancy things you have, from bath foam to deodorants, perfumes, and gels, and I'm none the worse for it!' An old lady standing beside us looked at us and nodded with a smile, as if she had heard this quarrel before. All of a sudden, as I listened to my own words, I realized they were not mine. I had heard them somewhere before, in politicians' speeches, at school, in textbooks. The same kind of argument, the same kind of logic ideologizing the past: we didn't have anything, but we still were happy. It was a lie, and I was participating in it. I took one more look at the bunch of coarse brown folded sheets, then put them back, sighing.

This was one more failed attempt to discipline my

daughter, to make her understand that – however crazy it sounds – we were not rich enough to buy toilet rolls, but only the packs of Golub, which were a third of the price. It didn't help to tell her that she was lucky there is paper to buy at all and that they don't have it in Poland or the Soviet Union, for example. She didn't consider this a good enough argument. All she remembers in her life is toilet rolls and she didn't see any good reason why, after seventeen years, she should have to give up something she considered perfectly normal. 'If this is a question of money, I'd rather not eat. But I don't intend to adapt to this kind of poverty.' Ashamed of myself, I finally took a package of the soft, pink rolls, because it was not only a question of money. My daughter was right; it was the principle of not giving in, not surrendering the basic elements of civilized living. I decided to defend the possibility of choice to the last roll of toilet paper, rather than allow communism to degrade our intimacy. But the old lady reached for the cheaper paper and went out.

This was in 1985. Even before 1989, people knew that communism was going to fail; they just thought it was going to take a hell of a long time. In fact, one of the indicators was toilet paper. So, when I saw Golub emerging again in the stores a couple of years ago, I thought, well, here we go, forward to the past. Through twenty years of stormy ideological, as well as some hygienic, changes in our lives, Golub toilet paper slept peacefully somewhere in warehouses and at the bottom of store shelves, hidden from our sight, giving us the illusion of real progress. Only the very old people, the pensioners, would buy it, and they had to ask for it, as if a modern store where one could buy imported pineapples or kiwi fruit was ashamed to display it beside the multicolored expensive rolls, the pride of the communist paper industry. But as communism moved backwards, Golub worked its way up onto the shelves.

My feelings were torn, and I usually compromised. When I received my monthly salary, I would buy rolls. Later on in the month, I would buy folded sheets for me and rolls for my daughter. After all, I was used to it and she wasn't and Golub *was* cheaper. But there was one thing that I couldn't escape while buying it: remembrance. I see myself as a child, sitting in a cold toilet, the walls painted with green oil paint. I am holding a rough piece of paper in my hand, smelling sauerkraut and beans (again!) from the kitchen, and looking at the top of my squeaky black Borovo rubber shoes, while one of the many tenants of our communal apartment is shouting outside the door: 'Hurry up, I know you're reading!' It was the poverty and deprivation that I remember, at a time when poverty didn't look terrible only because almost everybody else was equally poor – and it was considered just. But it was terrible for another reason, because we didn't even know that something better existed. As soon as we discovered that, and started to want better toilet paper for ourselves too, communism was doomed.

Perhaps this memory coincides with my going to school. It was only after I started going to school in the mid-fifties that I noticed that people kept newspapers in their toilets. At that time my mother would already send me shopping, and I learned what every single child living under communism had to learn, that you can't find everything you need all of the time, and most likely you can't ever find anything. For girls, this basic knowledge was even more important, since their future family's life depended upon knowing how to find things in spite of shortages. In postwar Yugoslavia, toilet tissue was obviously not a very important product, and it wasn't produced regularly (when it was produced at all), much less distributed. Toilet paper fell into an incredibly broad category of luxury items, such as furs, perfumes, gold rings, women's hats, gloves, or stockings, chocolate, candy, washing powder, or toys, even milk and meat – it all depended. The general rule was that anything at any time could be proclaimed a luxury.

Again, it wasn't the problem of building a factory for paper production (there were a few), but of understanding the necessity, the need for toilet paper. The revolutionary ex-partisan government didn't care much for it. The new leaders got it for themselves from so-called 'diplomatic stores,' where only high party members were entitled to a free ration of milk or chocolate or good French Cognac, for that matter. But who were these new leaders? Peasants, like most of the population of prewar Yugoslavia, more than 80 percent of it, in fact. And these people didn't care, either – after all, paper was paper, one was as good as another. They used newspapers: two pages of rough paper cut into pieces (to tell the truth, these newspapers were no more than party bulletins, so they fully deserved their fate). Today this could be praised, perhaps even explained away as ecological consciousness, but at that time it represented a lack of any kind of consciousness. In the requisitioned apartments they occupied in the cities, there was no toilet paper; usually there was just a nail with newsprint torn in pieces and hung on it. The newcomers kicked out the old 'bourgeois' tenants, taking over their luxury apartments, but not their 'luxury' habits. I hated it, because when I visited my friends who lived in such apartments, I had to put the pieces back together to try and read the comic strips. Frustrating; but they explained to me that cutting newspaper was *kulturno*, a cultural thing.

In my house, the newspapers *Borba, Komunist*, and *Vjesnik* were not cut, but left in the basket so everyone could read them. Of course, newspapers were used in times of necessity, but I remember that my father and mother stole notepaper from where they worked – there was typing on one side – so we used a thin, white onion-skin paper. It was hard-surfaced, it look a long time to go down, and I had to flush the toilet several times, but it definitely was better, smoother than newspaper. Not long after I finished the first or second grade, in the late fifties, Golub appeared – rough brown sheets, folded and pressed together in a square package, so when you took one, you pulled the next one too. It wasn't much better than newspaper. I didn't like it at all for another reason, its name. It was named for a bird (the pigeon), and even though at school I learned that paper is produced from trees, I was suspicious. I heard stories about soap cooked up from people in the concentration camp in Auschwitz – the stories about lampshades made of human skin, children baked in kitchen stoves, and so on, were a part of our lives then. We overheard them from the older people and then told them to each other. After all, too many witnesses were still alive, and we didn't have TV to distract us before sleeping – not even picture books. In my childish mind, one horror equaled another, and if one was possible, why not the other? Maybe Golub was made of pigeons – of feathers, of ground bones. Using it gave me an uneasy feeling, and I tried to avoid it; but as it was the only toilet paper available, I had to use it in spite of my resentment.

For the new kind of paper some sort of holder was needed, and soon enough we had an appropriate wooden box with a slit through which you could pull the leaves one by one. The box was good for another purpose, too: for holding cigarettes. In the school toilets, on the boxes, one could see the little burn marks that cigarettes left while their owners most probably were occupied with reading the literature not exactly recommended in the literature course. The problem was that these boxes were always empty and we had to use notebook paper. When I think about it now, I suspect pupils were stealing toilet paper from the school, if there ever was any. That would be logical, because if it was considered a luxury – and it was – it was worth stealing. But Vera, the friend who sat next to me in school, couldn't use her notebook paper. Her father forbade it, marking every page with a number. He was afraid she would cheat him and tear out a page with a bad grade. Vera had to borrow paper from me, a kind of intimacy that looked more like humiliation. But communism created this lack of any kind of privacy – in crowded communal apartments; in its morals, where everybody was comrade to everybody else; in the Communist Party, where every member watched over the 'correct' life of the others – because

71

only when there is no privacy can there be total control.

In the mid-1960s many things did change: there was a less centrally planned economy, more liberalism in politics, and a higher standard of living. That – as we learned only fifteen years later, after Tito's death – was based not on higher productivity, but on foreign credits that he took and had to give back. Progress in communism was marked by better and better quality toilet paper. Besides the inevitable basic Golub, the production of toilet paper in rolls started. This is where one could clearly see the social stratification. The communist state was pushing the concept of a classless society further and further into the future – so far that no-body could perceive it or truly believe in it anymore. And so, while we all still pretended to believe in the official ideology, in everyday life there were classes: the majority of poor people (Golub people); the less poor (the ones who managed to live in two-room apartments, with TV sets, appliances, and maybe a car, and who used toilet rolls); and the party/state functionaries, a class of its own. It was hard to glimpse into their houses, protected by high walls, watchmen, dogs, and a general element of fear.

I had such a chance only once, at the end of elementary school, when a friend of mine got sick. He was a nice, quiet boy, but he wasn't really a very popular person because he was driven to school in a black limo, thereby underlining the differences between us. The teacher sent me to his house to show him the new assignment. It was there that I discovered a kind of paper I had never seen before: thin, soft, double-sided, in a light blue color that matched the blue of the walls. On the back of the door a spray was hanging, and when you pulled the cord, the whole toilet would smell of a pine wood. 'Austria,' said my friend, when I asked him where it came from, as if it was obvious. This paper couldn't be compared with the rolls we were buying, because even if rolls were a giant step toward a better future, the paper in them was really only a little bit softer than Golub, which made one constantly aware of a system that neglected basic human needs. Too bad for the system.

It was only in the late seventies that toilet rolls became a normal thing, together with regular washing of hands, brushing of teeth, and bathing – but not using a deodorant. In other words, our hygienic habits slowly but surely changed for the better, and one didn't die immediately upon entering a crowded streetcar – only a little later. This was when Golub finally disappeared from the shelves, to- gether with the old wooden boxes. The boxes were replaced by new roll-holders made of metal or plastic, designed to match the color of the toilet seat, tiles, and bathtub (in new apartment buildings, the toilet was put into the bathroom, to save the space, I guess). This was the time of high standards of living and, consequently, hope for 'socialism with a human face.' The old bathrooms were redecorated, rusty metal bathtubs replaced by plastic ones. A modern, middle-class communist family – and every family wanted to fit into that category – had to have a newly designed bathroom, with every single item, from sink to toilet, all in the same color, including the towels – exactly as they saw it in the imported German interior design magazine *Schönen Wohnen* but that was not all. By then it was possible to buy perhaps ten kinds of shampoos and soaps, bath salts, body milk . . . for those who had money. Communism was coming into its luxurious state. Of course, one could argue here that the foreign paper was still softer (and had printed floral patterns) and that their soaps smelled better. But one had to admit we had come a long way, baby! And the final proof of progress was the fact that our toilet paper was even exported.

Nevertheless, there was one place where time stood still: in the public toilets. Whatever kind of holders they had, there still was no paper in them. It was here, in schools and restaurants, in coffeeshops and movie theaters, at train and bus stations, that one could judge how far our famous progress really had come. But the good, luxurious times didn't last long. In the mid-eighties shortages arrived, announcing the mortal illness of communism. There was no sugar, oil, electric power, coffee, toothpaste, detergent, not to mention fruit like oranges, bananas, or lemons. Toilet paper? It wasn't gone, but rolls were so expensive they slowly gave way to Golub, and people began to feel, as they had in the fifties, that they were lucky to have what they had – it could always be worse. Once back to Golub, it felt even rougher than before when it had made its first appearance and it was a sign of progress. Now it was an evident step backward. Thank God everyone was granted a passport, and Austria and Italy were only a couple of hours' drive away. It looked funny at the border: the dinar was so weak that people couldn't spend money on clothes or technical goods. Instead, trying to maintain certain minimum standards, they would buy packages and packages of toilet rolls, coffee, and detergent, and stuff their cars with it. And they were tired of poverty. Poverty is dirty, ugly, and it stinks. Communism is poor – therefore ... the government could manipulate the older generation, born before or right after the war. It was the younger generation that became its enemy. They simply were not ready to accept a deteriorating standard of living in the name of an ideology they didn't believe in, whose symbol was Golub. This was how the communists lost: when the first free elections came, in May 1990, the entire younger generation voted against Golub, against shortages, deprivation, double standards, and false promises. In the whole of Eastern Europe they didn't vote so much for democrats or Christians or liberals – or whatever the winning parties are called – as against the communists.

But democracy doesn't grant you toilet paper, rough or soft. In fact, it doesn't grant you any paper at all, at least in this part of the world. How am I to explain *that* to my daughter? I told her about my visit to democratic Poland, where my friend Ana has toilet paper only because she had stockpiled it in her cellar for years. 'I couldn't bear the thought of being without it, it was my obsession that I wouldn't have enough,' Ana had told me. I told her about my visit to the film industry workers club – an elite place in Sofia. It was after November 1989 and there was no paper in the toilet. I left the last roll I had taken on the trip to democratic Bulgaria for Katarina; she asked me to. In her bathroom, I had seen the same sheets of white notepaper I remember – which makes me fear that childhood under democracy in Eastern Europe will look very much like my own childhood ...

8

A Chat with my Censor

'I don't look like what I am,' said the man's voice on the telephone, a little nasal but pleasant. It was because of this sentence that I decided to meet with my censor, Comrade Inspector M. Officially – as he put it – he was from the state security police (SDB) in charge of the press. I knew such people existed, but even so, his sudden invitation for 'an informal meeting and chat' caught me by surprise. I could have refused, but something overcame my reluctance, drew me to the idea, and I agreed to see him the next day.

In Yugoslavia, the editor in chief, who has to be approved by the Central Committee of the Communist Party (among others), is usually also the censor of his newspaper. But the fact that he might lose his job if 'errors' appear in his publication is not considered a sufficient check. The SDB provides a special service for surveillance of the media. Comrade inspectors like M supervise how well the print press, TV, and radio follow the 'line' laid down by the party. Their job is also to watch editors and journalists closely, and to put pressure on them, more or less discreetly, when they think it necessary. It is not an easy job, because the line changes so often, depending upon which wing of the party – the more democratic or the more Stalinist one – is in power.

My first reaction was curiosity; I had never met with any SDB inspectors, perhaps because I was not important enough. Therefore, this was my opportunity to see the SDB's power personified in one man. Is he tall or short? How does he dress? What does he do with his hands while he talks to you? I guess this was a reaction to the literary tradition, to the long line of books describing KGB officers interrogating hundreds of thousands – no, millions – of people in Lubyanka, forcing them to admit to things they never did and then sending them to an icy death in Siberia. But it was unwise to think of Solzhenitsyn or Koestler, because my curiosity was quickly replaced by fear. Well, not really fear, but something very similar: thoughts about my possible sins. In fact, what had Comrade Inspector M said on the phone? Yes, he politely invited me to chat, but he also mentioned that he had something important to ask me, and this makes it very different.

But what could he ask me? I'm a journalist at a political magazine, and I'm not a member of the party (we don't even have to use the proper name of that party, the League of Communists of Yugoslavia here because there is no other). Perhaps my colleagues have been invited to such meetings too, but that hasn't been a topic among us – knowing this, Comrade Inspector M also told me that he hadn't called me at my office in order to protect me. I sometimes write articles that stir up public opinion and attract the attention of party leaders. But then again, everybody knows that is not what counts, because writing articles doesn't change a thing. Or does it? And which of the articles in the last issue could have attracted the attention of the SDB man enough to make him decide to call me? He might have spotted the piece on Albanians in the province of Kosovo, but it is also possible that he didn't like my recent article on cultural politics, about turning more and more toward Eastern Europe. It could be any or all of this. I was trying to figure out Comrade Inspector M's perspective on my 'errors'. Were they disturbing public opinion, by expressing unacceptable ideas imported from the West, introducing values foreign to our socialist self-management society, or spreading untruthful and dangerous information?

Or maybe I'm headed in the wrong direction. Maybe he's thinking about my private life – my hippie past, a couple of joints, hitchhiking through Europe, love with strangers. Then again, maybe he doesn't suspect me at all, but it has something to do with my first husband, who now lives in Canada. Or the second one, who emigrated to the United States. I have to admit that I'm traveling to the West much too often, that I do have connections there, that I speak several languages, that I subscribe to three American, two English, and one Italian magazine, and that I receive a lot of books and letters from abroad. My professors at the university were well-known philosophers and critics of the state. And besides, I'm a feminist. Sometimes they open my mail, and I have heard strange sounds on my phone more than once. Until this moment I didn't consider any of this important – or not important enough to worry about. I stick to the principle that the openness of my work is my best defense – which does seem a bit romantic now.

Across a small, round table in the café sat a short, thin man with a beard, wearing a black jacket – the type you might meet at a party but wouldn't recognize again. He looked like a secondary school teacher, there was some of that authority about him, and at the same time the clumsiness of the autodidact. At least he had told the truth: he really didn't look like a censor. Of course, the 'reason' for the meeting was banal, too banal to be true. He explained that a man had been jailed; in his address book was my name. He mentioned an unfamiliar, perhaps fictitious name. But the jailed man was a friend of a well-known enemy of the state and comrade inspector was wondering what possible connection I could have with him. I almost laughed with relief. I told him that my job and his were in some respects similar; we both needed connections and information. Therefore, I wouldn't be surprised if a paper with my name on it turned up in the pocket of a terrorist or of the French prime minister or Warren Beatty.

The conversation continued in the usual manner, 'usual' meaning that of two people on a train talking about the recent political situation, inflation, the danger of nationalism, the price of food. But we both knew there was something behind it. I was too cheerful and easygoing, trying to prove I had nothing to hide. He was too polite and charming, talking about his daughter, off to university this year, his mother, and his pet cat – after all, this was no interrogation. Yet, nervously smoking cigarette after cigarette, he looked somehow wasted, as if he would rather be somewhere else, doing something else. For one brief moment it seemed that he, not I, deserved pity. The fatal sentence, 'You are only doing your job,' was already on my lips when he said, 'You see, I'm not like some of my colleagues. I don't believe in crudeness. My opinion is that one should follow journalists' work from a distance, get to know them well. And then, if they repeat serious mistakes, warn them tenderly. It is usually enough: journalists are smart people.'

At the word 'tenderly,' I began to feel very uneasy. 'You know,' he continued, looking behind me, through the window of the café, 'in a certain way, we are friends. I know all of your articles, your books. And I know not only what but *how* you think, how you will react to certain issues. I must admit I only wanted to see you. You look much prettier than in your photos.'

I left the café and went to see my editor in chief. His office is a small, dark room, the walls covered with shelves and dusty books. A working table was cluttered with mail, papers, and empty cups. He listened carefully about my chat with the inspector. Then he leaned across the table and said in an unusually loud voice. 'You just go on writing as you always have, don't even think about that conversation. I would be the first to report it if there were any suspicious people or activities at this magazine, I sure would!' So saying, he pointed to the ceiling with his index finger. That is how I learned that he has a microphone hidden in his office. And then I understood perfectly the significance not only of censorship but of its subtler, deeper variations –

autocensorship, internalized in each of us, so that we don't need to chat with our censors too often, so that we make their job easier.

The conversation with Comrade Inspector M was absolutely unnecessary. What was important was the time between his call and our meeting, when I began to examine myself, to search for my errors, to look at my life through his eyes, to interrogate myself as my censor would. But I also understand that if he really needs to, he will find evidence even if it doesn't exist. The guilt I'm talking about is not a question of facts but of their interpretation.

9

The Strange Ability of Apartments to Divide and Multiply

Andrea is a thirty-year-old university professor working on her PhD in medieval ideas of autobiography. She lives with her father and his second wife in a two-room apartment. Her room is small, with no space between the desk and the bed. As she hasn't got her own office at the university – she shares one with two other colleagues – she has to work at home. Sometimes she gets depressed; there is no way for her to get an apartment of her own, even a small one. Her father doesn't have the money to buy one, the university won't give her one – besides, under the new government the old communist policy of providing workers with free apartments will disappear. It would still be fine if she could rent one, but she can't do this, either. The rent would cost her more than half her income, and the problem is that one can't live on the rest. Statistics show that 75 per cent of a family's budget is spent on food alone. The way out is to get married to someone luckier than she is. But more likely she will be stuck in the same apartment with her father; maybe she will even bring her husband in and have a child or two, all in the same space – that would be normal. Meanwhile, every time she gets depressed, she eats chocolate. Under the circumstances, I worry that she will get fat.

Alemka works as a journalist for a large news magazine. She is thirty-one and married, with a two-year-old daughter. She and her family live in a three-and-a-half-room apartment owned by her grandmother. Grandma occupies two rooms – she doesn't want to give up her comfort – her mother has one room, and the smallest room is a sleeping area for Alemka's family. As a working space they use the small entrance hall to the apartment. They, too, have no chance to buy or rent anything. Her husband is an unemployed professor of sociology; before he married Alemka, he lived with his aunt. It is not easy even to think about that, yet they will get more living space only after Grandma dies. 'We are like rats in a pot,' says Alemka. 'We fight and bite each other for no reason. That is, if you don't count the mere lack of space as a reason. Our family ties are sometimes too strong – I would like a little bit of loneliness, maybe even alienation.'

After she divorced, Maša, an economist, returned to her parents with her two children. She has lived with them for the last fifteen years in their three-room state-owned apartment. She sleeps in one room with her daughter, now seventeen; the other is occupied by her twenty-year-old son; and her parents live in the third. There was no other way for her. She could hardly have survived with the two small kids, one salary, and no financial help from her former husband. At least when the kids were small, they were safe with Grandma to do the cooking, take them to a playground, and see them off to school. But it also meant no privacy for her. Her whole life has been controlled by her parents. 'I don't remember the last time I dated a man,' she says bitterly.

Katalin lives in her three-room apartment in the middle of Budapest. After she divorced, her husband went back to his parents and she exchanged their small apartment in a new building on the outskirts of the city for a larger one in the old center that needed repair. She paid a substantial sum of money to an older couple for this transfer. As a translator in a publishing house, she didn't have that money, so she borrowed it from her father, a former diplomat. Of course, the whole transaction was illegal, and she still fears the possible consequences, even if the whole of Hungary – as well as the whole of Eastern Europe – is doing the very same thing, because this is one of the most popular ways to get more living space.

Jadwiga divorced her husband, but not her mother. After her father's death, she and her son came back to live in her mother's apartment. It is a tiny two-and-a-half-room place in a modern concrete and aluminum building on Sobieskiego Street in Warsaw. Her mother's room is the living room, her own room is the study, and her eighteen-year-old son sleeps in something that should be called a closet but here in Warsaw is still considered a room – perhaps because it has a window. Measurements are strange in Eastern Europe, invented for the practice of constricted living, and one has to adapt to them. Stepping inside, an outsider almost has to relearn how to breathe in rooms with low ceilings, crowded with furniture from the last century that doesn't belong here, with pictures and porcelain, dried flowers, books, the mask and satin gloves that an old lady wore at a gala party in the thirties, a soccer ball and sneakers – the souvenirs of three different lives going on in this tiny space that makes the air heavy. Jadwiga's ex-husband is now living in his mother's apartment, too – except that he was lucky. She died.

My friends Jaroslaw and Irena and their two sons live with Irena's mother, too. Perhaps space is not their biggest problem – they are in a spacious renovated three-room apartment in Prague's Old Town. It is the mother's presence that is so oppressive. She joins in their everyday life, insists on cooking her own meals, tells Irena that she spoils the boys or Jaroslaw that he is absent too much from the house – or asks why did they buy the boys a computer, they are far too young, and by the way, this meat needs more cooking, see it's red in the middle, raw, and we're not animals, are we? Irena, you shouldn't cut your hair so short, it doesn't suit you really, don't tell me that I didn't warn you in time, Tomasz, could you please, *please!* switch off that radio, I'll become deaf in this house, you don't respect an old person, yes, I know I'm old and useless, I'll be gone soon, but that's what you wanted ever since we started living together, isn't it, tell me, just tell me openly, I'm in your way, oh, I'll go to the asylum one of these days – in the meantime, Tomasz, *Stop that Idiotic Music Immediately*!

I could go on like this indefinitely. When I think of it, most of the people I know here in Yugoslavia, as well as in

Poland or Hungary, or Czechoslovakia, or Bulgaria, live the same way. The lack of apartments is such a common problem that after a while one simply doesn't notice it anymore. In fact, I have trouble recalling younger people or people of my generation who don't live like that, with their parents, even if they are past forty. Perhaps they inherited an apartment, or perhaps they are (or were) involved in politics – because this is the only sure way to get an apartment soon. Or perhaps they had a private business or were involved in the black market. Before the war, over 80 per cent of the population of our country lived in villages; it was an agrarian country. Now the proportion is just the opposite. When the great exodus to the cities began after the war, cities couldn't grow fast enough, houses couldn't be built fast enough, so people multiplied within existing apartments. What else were they expected to do – wait?

Twenty years ago, we used to live in a three-room apartment. There were six of us: my parents-in-law, my sister-in-law, my husband, our baby daughter, and me. Our thirteen by thirteen-foot room on the seventh floor of a skyscraper was our living room, study, sleeping room, and baby's room all in one. The baby's crib was next to our convertible couch, which we folded up during the day so we could move around. The room was buried in books, diapers, toys. We were both students, and we were supposed to study in that same room, because there were not enough facilities at the university. And we did, because to us it was not an unusual situation.

Before that I was living with my parents, my brother, and my grandma in a two-room apartment. In fact, it was a four-room apartment, but two of the post-revolutionary rooms were assigned to other tenants, a couple who had no connection with us. In the postwar and post-revolutionary environment, the government divided big apartments into rooms, forcing complete strangers to live in a kind of commune. We were lucky: these two people were silent, almost invisible. The only problem was the bathroom, located in 'their' part of the apartment; we had to be careful about when we used it. Taking a bath required extensive preparations, because the old-fashioned metal heater had to be heated with wood. Besides, it took forever. The common bathroom was also their kitchen, and most of the time the sink and even the bathtub were crowded with dishes. We established Saturday as our washing day (of course, under such conditions we couldn't do it more than once a week). Once preparations were done, the whole family took baths, and afterwards Grandma would do the laundry. Later on, we erected a wall of plywood between 'ours' and the 'tenants'' part of the apartment – of course, we needed to build our own bathroom, too. The wall was thin: one could hear coughing (the male tenant was a heavy smoker) and springs or floor boards squeaking. But it gave us the illusion of having our own, private territory, a life unwatched by strangers.

My first husband had an almost identical experience. Five of them were assigned two rooms; the third was occupied by another tenant. Joseph Brodsky describes the same situation in Leningrad in his essay 'In a Room and a Half':

After the Revolution, in accordance with the policy of

'densening up' the bourgeoisie, the *enfilade* [suite of rooms] was cut up into pieces, with one family per room. Walls were erected between rooms – at first of plywood. Subsequently, over the years, boards, bricks, and stucco would promote these partitions to the status of architectural norm. If there is an infinite aspect to space, it is not its expansion but its reductions. If only because the reduction of space, oddly enough, is always more coherent. It's better structured and has more names: a cell, a closet, a grave . . .

But perhaps this situation had some advantages, too. For example, a couple of years ago, in 1987, a serious sociological study was conducted at Split University. Professor Srdjan Vrcan was interested in a characteristic but illogical phenomenon: why, in spite of probably the highest unemployment rate in Europe and the fact that about 85 per cent of the unemployed were young, there is not any kind of social movement or protest against an economy that forces people to wait an average of three years for their first job. The results confirmed what was already suspected: the reason is the conservative role of the family in our communist society. A relationship that from the outside looks like a romantic tendency toward strong family ties in our culture has its less romantic side. Young unemployed people live in their parents' apartments; their parents feed them, dress them, even give them some pocket money. The family furnishes complete protection and, in fact, young people have no reason to protest. Besides, protesting wouldn't lead anywhere. The gigantic government bureaucracy – a system that was built just to keep communists in power and that perceives every spontaneous movement (whether for peace, ecology, or a simple demand for jobs) as a threat to their rule – would put an end to any protest very efficiently. They would be considered 'hooligans' and punished as such.

But the problem is that even when young people get jobs, they cannot get away from their parents and still need their support. In such a society – all over Eastern Europe and the Soviet Union – young people cannot move up into important positions. They remain at the very bottom, no matter how skilled. It might be called youth discrimination. Parents could be accused of infantilizing their children and prolonging the status quo, but at the same time they are keeping their children off the streets. How can parents divorce children – or children parents, for that matter? Later on it is, perhaps, too late; the grandchildren have come, and who but grandparents will take care of them? Who wants to expose kids to 'care' in kindergarten, to frequent illnesses, to early-morning dragging out of bed? Once again, when the children get a job and an apartment, divorce becomes impossible. To understand at least a little of this complicated situation, one has to know that this is a geriatric society, in which political leaders over sixty are considered to be 'still young,' not to mention those long dead, embalmed so they can live forever in their apartments – absurd mausoleums built of marble.

At the beginning, people multiplied in apartments. But later on, a strange phenomenon took place: apartments themselves started to divide and multiply. Like living organisms, prehistoric animals, protozoa perhaps, they divided into two or three, becoming smaller and smaller. Afterwards, with the help of a little money, they eventually grew bigger again. Someone who has not lived here cannot understand the feeling: to see an apartment divided in two might be the happiest moment in your life. When I married, I moved in with my husband's family. We lived together for eight years, then *his* father exchanged the apartment for two smaller ones, and we got a studio. His sister married, too, and moved over to her husband's mother's. In the meantime, my younger brother, living with my parents, got married and brought his wife to their apartment. After a while *my* father exchanged his apartment for two smaller ones. So, two three-room apartments, within ten years, had multiplied into four smaller ones. When I got divorced, my husband moved out to his new wife's apartment. I remarried (my new husband was living with his mother) and we exchanged the small apartment for a bigger one, giving some money to an old lady illegally so she would be willing to move to a smaller place (but with central heating!) Now we are once again in a three-room apartment, but my daughter is twenty-two and soon she will need an apartment of her own. What can I do but exchange my three-room apartment for two smaller ones, back to where I started from – a room and a half, 'if such a space unit makes any sense in English,' as Joseph Brodsky says. The historic progress in my case (and in the case of my country) is that I finally *can* divorce my child. But I wonder if anybody, except us here, could understand this strange process of the inorganic becoming organic.

Apartments for us were mythical objects of worship. They were life prizes, and we still regard them as such. Once you get one, it is all you can expect for the rest of your life. We seldom changed it, as we didn't change our job or the city where we lived. We were stuck with it, it became a part of our destiny, a reason for our fights and divorces, for our neuroses and fears of overcrowding, for our closeness with our parents and relatives. True, our babies profit from it. They have their grandmas to take care of them. But we suffer, thinking that if only we'd had our own place, life would turn around. More than that – an apartment was a metaphysical space, the only place where we felt a little bit more secure. It was a dark cave into which to withdraw from the omnipresent eyes of the state. I know it sounds silly, and I even feel embarrassed to think about it, but when I read Orwell's *1984*, the thing that frightened me the most was the TV screen that was also a camera, a spy. It was so chilling, and the fear stayed so deep within me, for years I refused to buy a TV, saying that I could do without it. And I could, because it had only one state-owned channel, brainwashing and boring us to death. But this was not my real reason, only an excuse.

Orwell exploited the uneasy feeling we all had. An apartment, however small, however crowded with people and things, kids and animals, is 'ours.' To survive, we had to divide the territory, to set a border between private and public. The state wants it all public – it can't see into our apartment, but it can tap our telephone, read our mail. We didn't give up: everything beyond the door was considered 'theirs.' They wanted to turn our apartments into public spaces, but we didn't buy that trick. What is public is of the enemy. So we hid in our pigeonholes, leaned on each other in spite of everything, and licked our wounds.

10

Our Little Stasi

According to a statement by an ex-KGB general,
Oleg Kalugin, it is permitted for everyone to be wire-
tapped, except the nomenklatura. *But people knew*
that *even before he did.*
— Argumenty i fakty *(1990)*

It is three minutes to 10:30 in the morning as I wait almost at the head of a long line in our tiny neighborhood post office. I am wondering if I will make it in time, or if I'll have to wait for the coffee break to be over. Behind the glass window the cashier accepts a telephone bill payment from a lady in a coquettish red hat who is in front of me. Then the cashier, who is a bleached blonde, gets up, puts up the sign, PAUSE, turns her back on the crowd, and starts making coffee on a small stove in the back. Too late – now about a dozen of us will have to wait a good half hour. 'Why does it always happen to me that the post office cashier shuts up in front of my nose?' comes a woman's voice down the line. But the others won't let her be the only victim. 'It happens to me, too, but I don't complain. After all, most of us are waiting here during our working hours, aren't we?' says a dry old man standing behind me. She shrugs. So what, isn't time still the cheapest thing in these parts of the world? Meanwhile, the lady in the red hat is counting her receipts. I can see that her telephone bill is 237 dinars. A smell of freshly boiled coffee fills the small space. A man's voice in the only telephone booth is shouting: 'No, no, Mother, I can't visit you this weekend!'

All this makes us feel somehow at home, as if the post office were a living room, and we had known each other for quite some time. Behind me, people are sighing. I don't only hear it, I can feel it on my neck because a fat man behind me keeps snorting. Even though I don't glance at it, I can see his hand with a money order for 450 dinars for his rent. Although I can't imagine what could interest me less at this moment, I almost automatically make a quick calculation: it has to be at least a two-room apartment, in a new building, because rents are cheaper in the old ones. Then I stop, ashamed of myself. The only reason I don't feel like a spy is that he too can 'spy' on me: he can see that I'm paying a 350-dinar installment for some books, and that my telephone bill is enormous, 1300 dinars. Perhaps right now he's wondering how I can afford such a huge bill, when my profession obviously has something to do with books, and we all very well know one can't live on any kind of intellectual work. In fact he really can learn a lot about my own and everyone else's lives just by waiting in a post office, bank, or any other institution that involves standing in line, paying bills, or acquiring money or some kind of documentation. But the post office is not a living room, and we don't know each other . . .

In November 1990, we citizens of the new democratic republic of Croatia received postcards with an unusual explanation from the central management of the Croatian Post Office and Telecommunications (CPT). You should know that citizens here are not used to receiving postcards from CPT with any kind of explanation, whether usual or unusual. They get only bills, warnings about unpaid bills, or additional bills for prices that have gone up in the meantime. The postcards were a double miracle because of what was written on them: from now on, when waiting in a post office, one must stand behind a yellow line on the floor. This yellow line will indicate a so-called 'space of privacy,' so that every citizen from now on will be able to do his or her business alone at the window, without someone constantly peeking over their shoulder. Not a big invention, one could say, since it exists in post offices, box offices, or banks all around the world. Yes, as we soon learned, it took the initiative of no less than our new president himself to bring this idea to life.

As he was an ordinary citizen before (well, not exactly; he was an ex-general), I imagine that he must sometimes have gone to the post office himself – although I bet most of the time it was his wife who went there. I also imagine him standing in a long line while someone breathes heavily behind him, pushing him back and forth. Perhaps he became sick of smelling garlic, sweat, and damp clothes, of looking at greasy hair or dandruff on the shoulders in front of him. Perhaps he asked himself, didn't these people ever hear of deodorant, toothpaste, or even soap? Surely he could easily imagine the germs of flu, pneumonia, meningitis, rabies, or other terrible diseases floating around freely – one doesn't need to be a hypochondriac to get paranoid, pressed up in a crowd against other human bodies. The president simply got tired of getting into other people's privacy – but particularly of other people intruding on *his* privacy, his money orders, his pension checks, his savings account. I am sure he and his wife grumbled about it at home, how indecent, how absolutely intimidating it was.

So, a few months after he became president, he remembered his lifelong experience with the post office. To be fair, we have to admit that he had other priorities, such as celebrating his victory, then more celebrations, dedicating a new monument on the central square, renaming streets and institutions, traveling to foreign countries, and so on. Considering all the problems of the republic – unemployment, riots by the Serbs, nationalism, the threat of a military coup – I'm quite impressed by how soon solving the post office problem came on to his agenda, unless there is a new presidential tactic involved that I was not aware of: to begin with a small change in post offices and banks, then progress toward the big, substantial, important changes in, let's say, politics or economics.

A story goes that the president was addressing the new city council. In his speech, he mentioned how we should change our rather primitive manners and suggested a red line in front of every post office or bank counter, 'so people would learn how to stand in line, not to step on each other's toes or give each other the flu, and to mind their manners as people in the world do.' It didn't take more than two weeks for the president of the city council to call a meeting of general managers of the post offices and banks, and for the postcards to arrive promptly at our doors. Except that the original communist red line in his speech was quickly changed to a yellow line. As expected, this move was saluted in the newspapers as a great leap forward, toward the already mythical 'West.'

All of a sudden, private space became important, even fashionable in a country where for forty-five years, if not longer, nobody had even thought in these terms, and it was perfectly normal not only to have to wait in line pressed tightly together, but to peer at each other's documents, accounts, letters, and bills quite shamelessly. Considering that privacy was a bad word, such peering was even safe. Asking for the right to privacy meant you had something to hide. And hiding something meant it was forbidden. If it was forbidden, it must have been against the state. Finally, if it was against the state, you must have been an enemy. Or at least a suspect person – the logic of the post office was basically the same as the logic of the state. But with that simple, if not original idea of the yellow or red line, the president introduced privacy as a public category, in this way transforming a mass of people into individual citizens. Or rather, he attempted to do so. For it takes more than a postcard, a speech, or even a decree to raise a mass, *narod* (a people), into individuals and to press the post office into service for these individuals – because the post office in any communist country is everything but a service institution.

The president's initiative is interesting in another respect, simply because it makes you think about privacy, about defining the meaning of the word itself. It makes you aware that you have a right to such a thing. It also makes you ask yourself, how come you forgot that privacy is normal? Except that the question itself is wrong, because how could you forget something you never learned or had? For me, something else was normal, for example, to see that my mail arrived open, especially mail from abroad, or to observe that letters from London, Stockholm, or New York sometimes take a week, sometimes a month, and sometimes never come at all. Due to bad weather? It is normal that I don't even ask myself how come my books or publications arrive all too often clumsily packed in a plastic bag with a ridiculous stamp, DAMAGED IN TRANSPORT. After all, that is a pretty correct explanation of what happened to it. I also never doubted that my telephone and the private phones of my friends were tapped – and as for the phone at my magazine, it was never even in question. We learned to recognize the signs of tapping: the phone rings, you pick up the receiver, and as soon as you start talking, there is that characteristic clicking sound on the line. Marko or Jozo or Ivan just plugged in. Hi, fella – we even used to salute the person on the other end of the line if we were in a good mood, continuing the conversation with a friend as if nothing had happened – or nearly so. But it was not much of a joke.

About the same time that the 'space of privacy' was introduced into the newly proclaimed democratic republic of Croatia, in the communist republic of Montenegro members of the Democratic Party discovered that in every post office building there is a special room for listening in and tapping telephone conversations, and only the police had keys to such rooms. The Democratic Party sued the director of the post office, as well as the Minister of the Interior and other responsible persons. The Workers' Council of the post office asked the police to return the keys. But the police authorities wouldn't even think of giving them back. Neither were they disturbed by the fact that their secret was finally discovered. Nobody offered their resignation. The Workers' Council, seeing no other way out, proposed a 'compromise': from now on, a policeman is not to go into the special room alone. He must be accompanied by a post office employee, and they will listen in together!

Not one of those accused denied the accusations. What else is one to believe, except that listening is not considered a crime at all – in other words, that it is normal? Even more interesting is that the whole case didn't attract public attention in Montenegro or in any of the other five republics, as if nobody can care less whether they are wiretapped or not. In spite of the fact that there is no reason whatsoever to believe that such a 'special' room doesn't exist in post offices in Serbia, Macedonia, or Croatia, no post office manager in any other republic, no Minister of the Interior, even from the most democratic republics of Slovenia or Croatia, would stand up and say publicly how ridiculous these accusations are or reassure people that it is not true because the post office is not a servant of the state's secret police – as if for both sides, citizens and officials, it is perfectly normal that the post office is one of the secret police departments, our little Stasi.

But what am I talking about? In the age of communications satellites, fiber-optic data transmission, modems, and so on, I am still considered lucky to have a simple telephone line. My own, mind you, not having to share it with some unknown person, as many of us are happy to do. (You know you are sharing a phone line when you pick up the receiver and there is no sound, it's 'dead.' It feels a bit like sharing an apartment with a ghost, apart from the fact that your ghost picks up the line just when you need it most.) To get a telephone number in Zagreb today, one has to pay approximately $2000 in a country where the average annual income is $5000. If there is an official explanation – which is unlikely, because CPT has a monopoly – it is usually that the exchanges are old and overworked. It is possible in theory to get a line on credit and pay it back to CPT in monthly installments, but even if you have the money to pay for it right away, you won't get your number. Even when it is in your contract that you'll get it in six months or, more likely, a year's time – you won't, unless you are a VIP or have very good connnections there. The ways of the post office – like those of God almighty – are mysterious, and nobody has been able to change that, not yet. Making a telephone line a privilege, rather than a necessity and the normal means of communication – that is where the power of the post office lies. More than a hundred years after Alexander Graham Bell sent his first message over the telephone wires, getting a number is a big achievement here. And should I even mention that the rental charge doesn't include providing a local directory?

If there is one thing that fascinated me in the States, it is the telephone system. After putting in an application, you get your number in three to four days; you pay, by our standards, an incredibly small sum of money ($30 or $40); you get itemized bills. And if you are not completely satis-

fied with your long-distance service, you can switch to
another company (what an incredible idea!), perhaps even a
cheaper one. There is more: when you call an official number, the line is almost always free (how do they manage
that?), even when you call information or the train station
or 911, all of which is unthinkable here. The culture of
communication goes so far that, if the person you are call-
ing is not there, someone will take a message and even deliver it – in case you have not left your message on an
answering machine, which is illegal here, if not tested by
the post office, and that costs dearly. And toll-free numbers
or collect calls sound like some kind of futuristic invention.

The point is that the American telephone system *works*.
My first impulse when I came back to my country was, sure
enough, to smash my telephone – which of course I didn't
do, because I remembered how precious (and expensive) it
is. But the proof that a 'telephone class' does indeed exist is
the latest gimmick that the post office has offered its desperate slaves-to-be. If you pay 100,000 dinars (about $8000)
in cash, you'll get a Mobitel or radio telephone, together
with a number, in an incredible eight days' time. No wonder the first 400 (out of a planned 10,000) sold out immediately. No other brand of radio telephone will be tested
or permitted. In the 'new democracy,' the good old post
office obviously has every intention of retaining its monopoly.

So, here we are, squeezed between these two concepts –
one of the post office as an accomplice of the state in the
sacred duty of protecting the famous 'security of the
country' from its enemies (and thus making communica-

tion more complicated, if not impossible), and the other of the post office as a service that facilitates communication. The problem with the first concept is that the history of communist states shows that the category of 'enemy' could spread to the whole nation. The problem with the second concept is that, in spite of good intentions or proclamations, we never have a chance to experience it. In some strange, twisted way, it was we who served the post office: we paid weird, unexplained bills; we never complained (and when we did, it didn't change anything); with our money we fed their arrogant clerks, and with our conversations we fed the police. The amalgam of fear and general helplessness on one side, and the need to be privileged, to be able to communicate, on the other, consecrated the post office, turning it into an impenetrable institution of power. In our state, however new and democratic it may be, the paranoia still goes on. And if a symbolic yellow line indicating the 'space of privacy' has to be drawn, it has to be drawn not only in front of the post office window, but behind it, too. Without that, what really has changed? Perhaps only the masters. Until they prove otherwise, I will salute the 'fellas' sitting in that little room and listening (I wonder why I think of the room as little?) and continue to tell my friends the code words: '. . . but this is not to be talked about over the phone.' Then, thanks to the post office, we'll meet in town to discuss it over a glass of strong red wine.

I have to admit that as a last resort I once or twice contemplated boycotting the post office, going back to carrier pigeons, for example. But I gave it up after my telephone line was out of order for just three days. Instead, I opted for outsmarting the post office. How? This I cannot tell you, not yet.

11

The Language of Soup

She is making noodles for a beef soup. Her fingers covered with flour, she puts the dough on the kitchen table, then rolls out a small pile with a rolling pin until it is thin, almost transparent. Then she lets it dry; later on, she will cut it with a knife into thin, long strips. She does it skillfully, and by looking at the pile one can tell it's going to make good, old-fashioned noodles: dark yellow, because it's made with yolks from fresh peasant eggs, not the anemic ones from the state-owned grocery store. 'I can easily buy readymade noodles at a store,' says Zsuzsa, 'but the kids like them best this way. There is no good beef soup without hand-made noodles.' Even if she is thirty-six years old and a working woman, Zsuzsa still makes soup and noodles like in the old days – the way her mother does. Or the way I do. She takes a pound or two of beef, a little bit greasy, adds a bone with marrow in order for the soup to be strong, and then cooks it up in cold water with parsley, celery, a carrot or two, and an onion cut in two and baked a little on the stove first. It has to cook for hours, until the meat becomes soft. Then she strains it and adds the noodles to boil for five to six minutes. At the end, when the soup is on the table, she sprinkles it with fresh parsley and a dash of pepper. The meat from the soup she cuts in thin slices and serves with potatoes and horseradish.

Her kitchen is a big, square, warm room, with black and white tiles on the floor and a window overlooking a busy street. A little untidy – there are some leftovers from the children's breakfast on one end of the table, unwashed pots with the grounds from yesterday's coffee in the sink. An old refrigerator roars, a radio plays American pop music. On a shelf above the stove I can see a box of Twinings Earl Grey tea, Nestlé's cocoa, Wasa crispbread, and soya sauce. She probably bought all this just around the corner, as in any West European city. Imports of goods into Hungary have been growing in the last few years, together with private in-itiative in the economy. But this kitchen is different from modern kitchens in Western European cities, because it is the central place in the apartment, the belly of it, the war-mest, most sheltered, peaceful and secure place, like a cave – the place that turns her apartment into a home. This is where her two kids come first when home from school, where she translates in the evenings, where her male friends come to smoke and drink strong coffee, or even stronger brandy, and discuss hard-core politics, and her women friends to smoke, drink, and discuss men and politics.

It's nine o'clock in the morning. Finishing the noodles, Zsuzsa washes her hands and sits across from me, sipping her Nescafé. The kitchen smells of the beef soup, simmering slowly on the gas stove. The familiarity of this setting, of the cold, foggy morning when, by some goddess's miracle, we don't have to go out to work, makes us both feel comfortable, even luxurious. The smell, the sounds from outside, her very face, makes me feel as if I have been here before. But I haven't, and that's why this feeling is strange and beautiful at the same time. Sitting in her kitchen, I sit in the middle of the world she has admitted me into because we share the same language – the language of soup.

I arrived this morning. A friend back home gave me Zsuzsa's phone number, I rang, and here I am. I see her for the first time in my life and we talk about men. How is a woman to tell the story of her life and not stumble upon men? Sitting at the table, she unfolds her past for me, as if she is unfolding a pile of dough. Zsuzsa and Istvan were very young when they married, just out of the university. He took a teaching job, she worked as a freelance translator. It went on for thirteen years – two kids, the peaceful existence of two intellectuals in communism who don't expect much from life, and in typical Central European fashion are content that it isn't getting any worse. It was Zsuzsa who finally, without any 'obvious' reason, decided that it would be better for her to get a divorce.

The two of them didn't fight. Of course, the kids stayed with her; this was not even in question. 'Most of the judges in Hungary are women, and we knew I would get custody,' she says. Her mother was against the divorce; she said it is better for a woman to have any husband than no husband at all. But Zsuzsa's father took her side. She got a job in a publishing house and another as a freelance journalist because she couldn't support herself and the kids on just one income. The poverty line is 4000 forints; she makes between 7000 and 8000. It would be stupid to ask her why she divorced, I thought, while she was telling me this. After all, why did I divorce? 'I felt so passive, as if life was just passing me by. I worked, took care of the children and the house, but I was in a state of deep hibernation,' says Zsuzsa. 'Something was missing, and it was a space for myself. When there is no space in society to express your individuality, the family becomes the only territory in which you can form it, exercise it, prove it, express it. But a family is too limiting, there is not space enough in it for self-expression either, and negative feelings accumulate very soon. We started to hate each other, but we stayed together because of the bigger enemy, waiting for each of us, out there – the solidarity of victims, I guess.'

For a brief moment her words – floating above the soup pot lid, the honking of the cars on the street, the music and distant voices – sounded almost like a forumla. How many times, in how many kitchens, over how many cups of coffee had I heard such a story of a woman struggling for herself – stories that we all know almost by heart? But it was something else that I detected, perhaps not so much in her words as in the very tone of her voice. The absence of anger, of fury, of ire, of the aggression and passion common to all of us when we are hurt. She was hurt, because his life was more fulfilled than hers, but beyond her anger I detected a softer, milder tone, a tone of deep ambivalence, as if I shouldn't take her words literally. For if I do, they will take on a crude, definite shape, become cold, solid objects. Estranged from her real feelings, her own words will turn against men, and this is what she doesn't want to happen.

I know this ambivalence. This, too, I have heard somewhere else in Eastern Europe. In a tiny kitchen in an apartment in a new concrete building in the outskirts of Warsaw, Danuta is telling me about her life with Tadeusz. We sit at her table, with space for only two chairs. 'This kitchen is so small, it makes me claustrophobic. When I was a kid in Kraków, we had a big kitchen with a wood-burning stove and a sofa next to it. That was a human kitchen, I'd say.' She too is cooking soup, but because it is hard to find a good piece of meat here, she makes a chicken soup, with a lot of vegetables, pieces of meat, and noodles. It's very thick, very yellow; we call it *ujušak* or *eingemachtes* in Zagreb. I can see that she wants to talk about her feelings after almost twenty years of marriage to Tadeusz. But before she knows it, she starts to speak about *his* fate, the fact that he had to leave the Academy of Arts in 1968 as a dissident. Later on, he was expelled from the film studio where he worked as a documentary film director. She supported him by working as a sociologist in some state institute, so at least one person in the family held a steady job. In the meantime, he worked at odd jobs, wrote poems, published them, and became well known. It is now about three years since their divorce. At Sunday lunch, which he comes to every week so their son can feel that at least in this way they are still a family, I see a timid man with wire-rimmed glasses, who talks about the Polish economic crisis and how his organization, Solidarity, has changed for the worse since coming to power. He looks tired and deeply disappointed. At forty-six, he feels that everything is behind him. His most creative years he spent struggling with the pressures of the system, just trying to survive. He is a victim, and he makes this clear.

Even if she doesn't say so, Danuta is no less a victim. She had to work to keep her family going – someone had to do it. And she didn't have time to think of what she really wanted from her life. 'Then, as I woke up one day, I realized that our son had grown up, that my husband has a young and beautiful lover, and that I am forty-eight and alone,' she says calmly, as if what happened to her is not a matter of Tadeusz or herself but of destiny. They divorced, but it was too late for her; she felt betrayed by him, by life itself. She is not bitter, just empty. And she doesn't blame him for what happened – not even for running off with an actress three years older than their son.

At that lunch, looking at Danuta, I realize that we don't blame men easily, as if we don't have the heart to do so, even if it would be good for our self-esteem. Zsuzsa and Danuta equally hesitate to place the guilt on their men because, as they say, 'we all live in the same mess.' It would be so easy to put the blame on them; our lives would be so much simpler. But something stops us. Maybe it's the idea that in normal circumstances, things might have been different. If only we could live normally – it doesn't really matter what this 'normally' means, as we perceive our own situation as abnormal. Then perhaps we would discover that our plea for normality is only an excuse for our own inadequacy and an inability to deal with our problems. Or maybe it is a question of having a weak character that is our problem, and ours alone, and society has nothing to do with it. But if so, we are not in a position to become aware of it. There is always this strange feeling that this is not life as it should be, or could be . . .

Veronika's kitchen window looks out on to the backyard. It is in the old part of Prague – Stare Mjesto – where the houses look like fortresses, with strong walls and small windows. The facade across from her is crumbling from damp. The window is wet with raindrops. In the semi-darkness of the kitchen, Veronika moves silently, slowly. She doesn't have to tell me anything – her story, I mean. I can easily read it in her attitude, her movements, her absentminded glancing look, and her silence. She is making farina dumplings for the beef soup: one whole egg, a pinch of salt, a little oil, and a cup of farina. She mixes it automatically (the secret of good dumplings is in the mixing); there is no need to concentrate on her movements, this is what she does every Saturday. Nobody is home, her two sons are out, her husband Jiri, too. I know it's him that she is thinking about - his absence, longer and longer every day. She doesn't want him to leave, even if she knows he has a lover.

She doesn't have time during the week to cook; the kids eat at school, she and Jiri too. So on Saturday she goes to the open market and buys food, fresh meat or fish, maybe vegetables, depending on what there is to buy. Today she found kiwi for dessert. It was very expensive, but she bought it because it is Saturday and they will all eat together. For Saturday lunch she cooks meals that the kids like, that he likes . . . as if she is mixing some kind of magic potion that will keep her family together – at least for a while, at least around the table. It takes a long time for the soup to cook, but finally she speaks to me. 'It's just that he is so unhappy, so frustrated. I understand him. In such a system I am happy to work as a little bureaucrat. But he is a journalist, and you know what kind of pressure they suffer. I don't know how I can help him – or us. Because if he leaves us, I feel like everything will fall apart, even these thick walls around me.' After the long silence, her words fall heavily, like the drops of rain outside.

In these endless, ancient, kitchens, while the soup is boiling, we talk of men – sorrowful talk, as old as the smell of soup, as warm as a fireplace and the feeling of intimacy we share. It seems it's the same the world over, but there is a difference here, because every such talk finishes with the system that shapes our lives. Sometimes it looks like a Minotaur, a monster we throw sacrifices to in order just to stay alive, a force larger than life, mythological and real at the same time, controlling us, dividing us, eating us up. That is what is uniting us in these kitchens, beyond the men, above them, this feeling of helplessness. And this is also what unites us with men. It's hard to see them as an opposite force, men as a gender, hard to confront them as enemies. Perhaps because everyone's identity is denied, we want to see them as persons, not as a group, or a category, or a mass. A blond, slim man playing football with his sons on Saturday morning and later sitting opposite us, nervously rubbing his hands; a man who cannot find a decent job for twenty years. A weak yet perhaps cruel man, who speaks softly, and his poetry is superb . . . I take Danuta's or Zsuzas's or Veronika's side – that is my side, too. Yet, at the same time, all of us are able to take Tadeusz's or Jiri's side, and this confuses us, because we are not able entirely to distinguish *us* from *them*, and all of us together form *it*. So, in our kitchens, while the soup is boiling, what we talk about is identity.

12

A Communist Eye, or What did I see in New York?

Just how many 'iron curtains' still exist? What are they made of? Sometimes it seems to me they are made of a material stronger than iron itself: our memories.

It happened one night last summer in New York City. I was coming out of a train; there were not many people in the subway. As I approached the steps, I spotted a muffin lying there – a whole big, brown muffin, nicely wrapped in plastic, in the middle of the steps. As if somebody dropped it in a hurry that very moment. And sooner than I could even think of it, I reached out my hand to pick it up. Or rather, my hand, my spine, my whole body bent toward it, without my knowledge or command. At that moment I was not hungry – in fact, I was just coming from a dinner – but I was not conscious of what I was doing. Halfway down, I stopped. I didn't pick it up. I grabbed for the railing and stood there for a while, surprised at myself. Then a wave of nausea overwhelmed me: *Why did I do that?* What in the world was the reason to make such a move? Where did that come from?

I went out onto the street, still thinking of the muffin on the steps. Right by the exit there was a blue trash can, and as if the muffin had set something moving in me, I compulsively stopped in front of it. There were some newspapers, empty boxes, bottles and plastic glasses, but most of it was food: a half-eaten Chinese meal in a polystyrene box (even the plastic fork was there), an open bag of potato chips, a few apples (one could say they were rotten, but it depends what your standards are), part of a McDonald's hamburger, a piece of banana, a slice of pizza – all in just one trash can. I could not help myself; it fascinated me, all that food thrown away. I needed to remember where I was and that, unless I was a beggar myself, I was looking at something called 'garbage,' not exactly food, at least, not here.

A block from the subway entrance there is a church. As I was passing by, I saw a man sleeping in the doorway with cardboard underneath him, instead of sheets – a homeless person. Not that I hadn't seen him before, and many others too, and that they hadn't given me an uneasy feeling, only now I became somehow more intensely aware of him. In two months I haven't got used to it, I thought as I passed him. I just can't. This is what all my American friends were telling me, 'You'll get used to it, just wait and see. After a few days, you won't even notice them – homeless, beggars, underclass people.' But every day that I spend here I become more and more conscious of them – as if, walking down the New York streets, at some point I see *only* them and cannot take it anymore.

That woman, for example. She is sitting on the sidewalk in front of an antiques shop on Lexington Avenue, between Seventy-fourth and Seventy-fifth Streets. Perhaps she is my age, perhaps even younger, it's hard to tell. Dressed in black pants and a T-shirt she would look ordinary, if it weren't for her strange, worn-out face and tangled hair. Beside her there is a piece of paper saying 'HOMELESS' and a paper cup with coins. No explanation; nothing. Who would have time to read it, anyway? Maybe she isn't really homeless, maybe she drinks or spends money on crack. But there she is, every day, begging. Or the young man in the subway, covered with a blanket. 'I have AIDS,' says his sign. Maybe he does; I assume he does. In the subway, a black man with a walking stick is shouting, 'I'm a human being, too!' Never before had I heard a person saying such a thing. Terrified at the fact that he actually had to *shout* it, being forced to witness his humiliation, I gave him money. It was the least I could do. Not much, a quarter, because when you come from Yugoslavia or any Eastern European country, your money doesn't go very far. But it didn't help me to forget him. I wasn't able to buy my peace of mind.

I lived with a friend on the Upper East Side. Even there, in that elite part of Manhattan, you met one beggar every five to six minutes, no matter if you walked or rode the subway. I tried to select people; for example, to give money only to women and invalids, who are underprivileged even among the street population. I tried to play God, deciding who was more worthy of my few measly coins. But soon I gave up; it was not worth it. Perhaps God has more insight into their lives and destinies, but we human beings certainly don't. It's phony to make such choices because there is no proof you are right. A woman might lie when she says she is homeless, she might give her money to her drunken husband who beats her up. Then, again, she might not, but how am I to know? There are fake blind men – people even make a business out of it. But all in all, even if they 'lie,' I doubt if it's an easy job for any of them. For myself, I concluded, it's easier to give than not to give. It's easier to believe them. Then I don't have to fight my conscience, prove to myself that I don't need to give, which seems to be as hard as to make sure you gave money to the 'right' person – because there are no 'right' people.

There, in New York, I learned to give. The poverty is so vast one just can't keep one's eyes shut. At home it looks different, it involves another kind of dilemma. First of all, there are not so many beggars, though one becomes aware of that fact only when one can compare. Until a few years ago, the only beggars were Gypsies, and they were treated as if their begging were normal, as if it were their real profession – the state gave up on 'civilizing' them long ago, it would have been too much trouble. But in the last two years, with the severe economic crisis, all of a sudden a different kind of beggar has come to light. There is an old man at the corner of Bakaceva ulica, near the cathedral. Everyone calls him 'Professor.' He comes on Saturday mornings, neatly dressed in an old-fashioned three-piece suit, opens a folding stool, and takes out his violin. Then he plays for an hour or so and, occasionally, people give him money. 'How are you today, Professor?' someone asks, and he replies, 'Thank you, I'm very well, for my age, you know.' I heard he really is an ex-professor of music, over eighty, and that his pension is too small to survive on.

Then there is the woman who plays an accordion near the market. She doesn't exactly play it; rather, she sits there, running her fingers over the keys. The sound is not important, it's the work that counts, the attempt to make it appear to be work, not begging. It makes it easier for them and for us. Most of them do get some money from the state, but not enough to survive on. They are not young, not homeless; mostly they are old ladies and handicapped people, the orphans of a socialist system giving way at the seams, falling apart. But it took us – citizens – some time to realize that they are begging and that there is nobody else to give to them anymore, except us.

Caught between two sets of values, one where beggars are not allowed at all, and the other where beggars are the consequence of capitalism, we simply are not sure how to deal with them. There are institutions, but they don't function. Social workers give up, because in spite of good programs, they have no funds. Judges give up, because there actually is no other way for these people to survive, and they are relatively few. We hesitate because we are not supposed to see them, they are not supposed to be there. In the categories of socialist law, there is no begging, only 'parasitism.' In the categories of the socialist morality, there is no charity. So, if officially there are not beggars, officially there is no charity. And didn't they teach us at school that 'social parasitism' is one of the worst evils of society? In 'capitalist' countries, it is simpler, beggars are a sign of social injustice, class division, lack of social concern. The difference is that people there never even had the illusion we were brought up with, that they are all equal and that 'in socialism, the human being has the highest value of all.' On the other hand, when more than half of the working population is living at the edge of poverty, it is only a miracle (no Westerner can understand it) that there are not more beggars. In a situation where socialist ideals were so obviously falling apart, with a decaying economy – and when poor people couldn't work even if they wanted to, because of an unemployment rate that topped 14 per cent – to beg was quite simply moral.

In New York, I met my friend Evelina from Bulgaria. 'Imagine,' she told me, 'only six months ago I would never, but never, have thought that I'd see New York.' Apart from the excitement of being out of the country for the first time, though, she was depressed. One day, when we were walking in the Village, she said: 'Did you notice how many poor people there are? I just can't help looking at them. And even if I don't live here and don't have any personal connection with them, I feel guilty all the time.' This was it; we were seeing the same thing. But why? Why did we – as it seems – notice only the poor people, the garbage cans full of food, the litter, and the dirt? Did someone promise us something else? And if so, who? 'I think it comes from the movies and TV,' Evelina said. 'This is how we first learned about the States, through pictures. But beautiful, clean pictures. There was no dirt in them and no real poverty without a happy ending. We made the mistake of taking them for real. Have you ever noticed how the colors in movies are so very different, so much brighter than in reality? I guess I'm just disappointed that the product, life, doesn't correspond to the advertisements we've seen. Perhaps we should file a complaint with President Bush, what do you think?' said Evelina, laughing at this very American idea of filing a complaint.

Not only the colors but the people look different, too, in TV ads especially. I could watch them for hours, like a long soap opera from another planet. Yet on the street you don't see them, these polished, shining, smiling, perfect people. They are nowhere, and what you do see is much more average and much more scary. So – yes, we are bound to be disappointed, because we can't make a clear distinction between the picture and the reality, and because it is the promise that counts, the illusion that somewhere out there they do exist, only we don't see them. Then, again, maybe we just can't see them, maybe we have a certain kind of inborn blindness. There has to be more to it, there has to be another reason why we, Evelina and I, notice only certain things and certain people. There is a deeper reason why the poverty sticks to us, why we recognize beggars, homeless people, bums, petty thieves, drunks, the sick, junkies, why we take it all so personally, why it hurts us. It's because we have a *communist eye*. Like a third, spiritual eye placed in the middle of one's forehead, this eye scans only a certain type of phenomenon; it is selective for injustice. Even if the socialist states have fallen apart, the ideals of equality and justice haven't. They are still with us, built in like a chip. We remember them from school, from our movies, from literature glorifying the idea of justice, as well as from the clean, beggarless streets of our cities.

Ina, my friend from East Germany, told me that on her very first shopping trip to West Berlin, she took her kids, twelve-year-old Erna and nine-year-old Peter. With them, it was an arduous emotional experience. 'Of course they were fascinated with the toys, the food, the clothes. Who wouldn't be?' she said. 'But as soon as we started walking down the Kurfurstendamm, especially near the famous Zoo subway stop, I could see their consternation. They had never seen so many people begging, drunk, lost, or just sick, obviously needing help. They asked me why. What was I supposed to tell them, that the state that just collapsed wasn't so bad, after all?' Now Ina's kids and she herself have to 'get used to it,' but I, and she too, suspect it won't go very quickly and without consequences. One doesn't lose one's third, communist eye that easily.

Evelina and I see New York as Erna and Peter see West Berlin, with the memory of our cities still inside us. We have no other yardstick, no other way to see it, but against the situation we came from. That's why it is absurd to tell us that we should forget. Maybe there is nothing to forget, maybe this is something that we care to preserve, the idea of justice from the other world that we brought with us. We don't know yet, we need time to decide. Transplanted to the United States, we carry that idea and much more with us, like excess baggage that perhaps we would like to drop off or leave at the entrance to this other, promised world.

But even to look at the richness of a consumer society becomes difficult; we are not used to that, either. At Bloomingdale's, after an hour of touring the store, Evelina

stops for a moment as if she is gasping for air. 'I'm tired,'
she says. 'My head is spinning and my eyes hurt. There's so
much light here.' Even though this is not my first visit, I
feel exhausted in the same way. After a certain point, my
eyes refuse to look, my mouth becomes dry, and I start to
have a headache. I recognize this particular tiredness, hers
and mine, the feeling that it is just absurd to look at so
many things and so many kinds of one thing, as if one is
enclosed in a room with mirrored walls that endlessly reflect each other. It has to stop somewhere – you think –
this multiplying, this plenitude doesn't make any sense.
Coming from the world of shortages, one's idea of plenty is
mainly of fruit, meat, vegetables, of shampoo, soap, or toilet
paper. Here, you are murdered by variations on each of
these and by the impossibility of distinguishing the differences. First you discover an immense greed, a kind of fever,
a wish to buy everything – the primordial hunger of consumerism. Then you discover powerlessness – and the very
essence of it, poverty. Moreover, you start to realize that
Bloomingdale's for you is a museum, not a real store where
you can buy real things for your real self.

In socialism, we were not used to thinking of ourselves
as poor; the communist principle of *uravnilovka* (leveling)
made us all live more or less under the same conditions.
There were no ways, no means, not enough goods to estab-
lish a real, visible, palpable class distinction between poor
and rich. To be rich was an exception from the general
rule. The good thing was that we were all poor; the bad
thing was we didn't know it. This is what the Iron Curtain
is made of: many facets of a different reality, of different

ideals and meanings we were brought up with and truly believed in. The Iron Curtain is hidden in the feeling Evelina and I have as we walk down Fifth Avenue, knowing we are here and, at the same time, are not here because this reality has another meaning for us. In the middle of New York we are enclosed by a different reality principle. It's good to come here and see; for a prisoner, it is worth it for the sake of traveling itself. But it becomes clear that the instant 'translation' into another culture, into another way of life and values – and that is what people in Eastern Europe expect to happen – is impossible. The 'iron curtains' will stay with us for a long time: in our memories, in our lives that we cannot renounce, no matter how difficult they were and how hard we try.

13

A Letter from the United States –
The Critical Theory Approach

'**D**ear *Slavenka*,' her letter began; a two-page, single-spaced letter written on a computer. (By the way, I don't remember receiving a handwritten letter from the United States in the last couple of years.) '*I am writing to you about the interview I did with you in New York, in April, right after the Socialist Scholars' Conference (in a luncheonette near Gloria Steinem's apartment, if you remember) . . .*'

I remember – indeed I do. We were sitting on red plastic chairs, leaning over a plastic table, holding plastic cups with insipid American coffee, and B asked me about the position of women in Eastern Europe after the 'velvet revolution.' I also remember a kind of geographical map appearing in my mind: Poland, Czechoslovakia, East Germany, Hungary, Bulgaria, Romania, Yugoslavia too – we are talking about perhaps 70 million women there, living in different regions and cultures, speaking different languages, yet all reduced to a common denominator, the system they were living under.

It was after I spoke at the plenary session at that conference that B approached me. The big midtown auditorium at CUNY was almost filled. I was to give a paper on the same subject: women in Eastern Europe. But before I started my speech, I took out one sanitary napkin and one Tampax and, holding them high in the air, I showed them to the audience. 'I have just come from Bulgaria,' I said, 'and believe me, women there don't have either napkins or Tampaxes – they never had them, in fact. Nor do women in Poland, or Czechoslovakia, much less in the Soviet Union or Romania. This I hold as one of the proofs of why communism failed, because in the seventy years of its existence it couldn't fulfil the basic needs of half the population.'

The audience were startled at first; they hadn't expected this, not at a scholarly conference where one could expect theories, analyses, conclusions – words, words, words. Then people started applauding. For me, the sight of a sanitary napkin and a Tampax was a necessary precondition for understanding what we are talking about: not the generally known fact that women wait in long lines for food or that they don't have washing machines – one could read about this in *Time* or *Newsweek* – but that besides all the hardship of living in Eastern Europe, if they can't find gauze or absorbent cotton, they have to wash bloody cloth pads every month, again and again, as their mothers and grandmothers and great-grandmothers did hundreds of years ago. For them, communism has changed nothing in that respect.

But I wasn't sure that my audience grasped this fact, after all: first, because they were mostly men and, by some caprice of Mother Nature, men usually don't have to wash bloody cloth pads every month; second, because they were leftists. I know them, the American men (and women) of the left. Talking to them always makes me feel like the worst kind of dissident, a right-wing freak (or a Republican, at best), even if I consider myself an honest social democrat. For every mild criticism of life in the system I have been living under for the last forty years they look at me suspiciously, as if I were a CIA agent (while my folks, communists back home, never had any doubts about it – perhaps this is the key difference between Eastern and Western comrades?) But one can hardly blame them. It is not the knowledge about communism that they lack – I am quite sure they know all about it – it's the experience of living under such conditions. So, while I am speaking from 'within' the system itself, they are explaining it to me from without. I do not want to claim that you have to be a hen to lay an egg, only that a certain disagreement between these two starting positions is normal. But they don't go for that; they need to be right. They see reality in schemes, in broad historical outlines, the same as their brothers in the East do. I love to hear their great speeches or read their long analyses after brief visits to our poor countries, where they meet with the best minds the establishment can offer (probably speaking English!) I love the way they get surprised or angry when the food is too greasy, there is no hot water in their hotel, they can't buy Alka Seltzer or aspirin, or their plane is late. But best of all I love the innocence of their questions. Sitting in that luncheonette on Seventy-fifth Street with B, I resented the questions she asked me, the way she asked them, as if she didn't understand that menstrual pads and Tampax are both a metaphor for the system and the reality of women living in Eastern Europe. Or as if she herself were not a woman – slim, tall, smart-looking and, surprisingly, dressed with style. Feeling the slick plastic cup in my hand, it came to my mind that her questions are like that – cold, artificial, slippery, not touching my reality.

'*I am sorry to have taken so long to get in touch with you. I was in Berlin for a while this summer,*' the letter continued. '*I am doing a bigger project now on women and Eastern Europe – trying to put together an anthology on this topic. There is already a publisher who has expressed interest. I hope it will be more than a description of events, but some kind of analysis about women and democracy, the public sphere, civil society, modernization, etc. A kind of Critical Theory approach . . .*'

I picked up this letter from my mailbox on my way to the office (together with an American Express bill, which I didn't want to open right away because I knew it would upset me). 'She spent several weeks in Berlin,' I thought, reading it in the streetcar, 'and here she is, making an anthology!' How easy, how incredibly easy it is for her; she even has an editor. Women in Eastern Europe hardly existed as a topic, especially for leftists. And now, what is wanted is no less than a Critical Theory approach! I admit this letter upset me much more than the American Express bill would have. Following her instructions, I am to write '*some article specifically on women in Yugoslavia, dealing with the kinds of interventions women have made in the public discourse, eg, about abortion, women's control over women's bodies, what sorts of influence women have had in the public discourse on these topics,*'

126

and what sorts of influence the non-feminist media have had on women's issues now.'

Reading all this, I couldn't help laughing out loud. A few people turned their heads in surprise, but I didn't stop laughing. Women's influence in the public discourse? For God's sake, what does she mean? There is hardly any public discourse, except the one about politics. Women don't have any influence, they barely even have a voice. All media are non-feminist, there are no feminist media. All that we could talk about is the *absence* of influence, of voice, of debate, of a feminist movement. *'Do the women in Yugoslavia argue for an 'essentialism,' ie, that women are different from men, or is it a matter of choice?'* I read in her letter, with utter amazement. With each of her words, the United States receded further and further, almost disappearing from my horizon. Argue what? Argue where? Somehow, in spite of her good intentions, I felt trapped by this letter, the views she expressed in it, like a white mouse in an experimental laboratory. Sitting in her office at the university, with a shelf full of books on Marxism, feminism, or Critical Theory within reach, B asks me about discussion on *'essentialism'* in *Yugoslavia.* I can imagine her, in her worn-out jeans and fashionable T-shirt, with her trimmed black hair, looking younger than she is (aerobics, macrobiotics), sitting at her computer and typing this letter, these very words that – when I read them in a streetcar in Zagreb ten days later – sound so absurd that I laugh even more, as if I were reading some very good news. *'No, dear B, we don't discuss this matter,'* I will answer in my letter. *'It is not a matter of choice, it is simply not a matter at all, see? And I cannot answer your questions, because they are all wrong.'*

But if she doesn't understand us, who will? What is the way to show her what our life – the life of women and feminists – looks like? Maybe instead of answers, I could offer her something else. Suppose that my mind is an album of myriads of pictures, photos, images, paintings, snapshots, collages. And suppose I could show her some of them . . .

It is the autumn of 1978 and eight of us are sitting in Rada's room on Victims of Fascism Square in Zagreb. It is a little chilly because a balcony door is open, but it has to be that way. Rada hates smoking and yet we all smoke in the excitement – even Rada herself. This is because we have just come back from Beograd, from the first international feminist conference, 'Comrade Woman,' where we met the well-known Western European feminists Alice Schwarzer, Christine Delphy, and Dacia Maraini for the first time. We thought they were too radical when they told us that they were harassed by men on our streets. We don't even notice it, we said. Or when they talked about wearing high-heeled shoes as a sign of women's subordination. We didn't see it quite like that; we wore such shoes and even loved them. I remember how we gossiped about their greasy hair, no bra, no make-up. But all that didn't stop us from deciding to form our own group, the first feminist group in Yugoslavia. We didn't know how to organize; it even seemed impossible. First we talked. Then we published some articles – nothing big, of course. In a matter of days we were attacked by the official women's organization, Women's Conference, by politicians, university professors, famous columnists –

for importing foreign ideology.

So we discover that a feminist is not only a man-eater here, she is an enemy of the state. Some of us received threatening letters. Some got divorced, accused of neglecting their families. A maniac broke into my friend's apartment (convinced that he could understand her!); a writer wrote a porno story about two of us, feminists. Women themselves accused us of being elitist. A man wanted to chain me in the main square; someone spat on my door every night, for years . . . On the other hand, more and more women were joining in, attending our monthly meetings, participating in discussions, forming their own groups – a hundred, perhaps, at the beginning. But it was lonely being one of a few feminists twelve years ago. Sitting in Rada's room and making plans, it's good that we didn't know it then.

Twelve years later, when I was in Warsaw in 1990, Jola took me to another similar room. In fact, it looked like a replica of Rada's, even if this one was not in an ambassador's apartment, with original paintings, antiques, and Ming vases all around. It was in a skyscraper somewhere on the outskirts of town, but the atmosphere was the same: nine young women and their expectations. I ask them why they joined the group. One – a teacher, tall, married, no children – answers jokingly: 'Because my husband always interrupts when I talk. It's hard to recognize discrimination when you live with it.' They don't know how to organize yet, but they do know that feminism is about prejudices, about woman's self. Three of them had already participated in organizing the first demonstration against the anti-abortion law proposed in the Polish Parliament in May 1989. One came for the first time this very evening. When you think about feminism in Poland, you can count the women on your fingers: Ana, Malgorzata, Stanka, Barbara, Renata, people in this room. 'You might laugh at us, but we *are* the Polish Feminist Union,' says Jola. 'It's hard. Women don't take the initiative here; they wait for somebody to solve their problems – that's very typical for Polish women.'

That evening, in her apartment, still in Warsaw, Ana takes down a book from her shelf – a rather thick, ordinary paperback. It looks old, because it's worn out and somehow shabby. But it's not ordinary. I can tell by the way she handles it so carefully, like something unique. 'This is the book I told you about,' she says, holding out the *Anthology of Feminist Texts*, a collection of early American feminist essays, 'the only feminist book translated into the Polish language,' the only such book to turn to when you are sick and tired of reading about man-eater/man-killer feminists from the West, I think, looking at it, imagining how many women have read this one copy. 'Sometimes I feel like I live on Jupiter, among Jupiterians, and then one day, quite by chance, I discover that I belong to another species. And I discover it in this book. Isn't that wonderful?'

She reminds me of Klara. In Klara's bedroom in Budapest there is a small shelf with about twenty such books. She has collected more because she is an English translator, and she travels to London from time to time. 'I read these books when I'm tired and depressed from my everyday life, from the struggle to survive and keep my head above water in spite of everything. Then I just close the door – leaving my job, two kids, the high prices, outside, no men – and read Kate Millett, Betty Friedan, Susan Brownmiller. It's like reading science fiction, an escape from reality. It's so diffcult to be a woman here.'

I see that when I visit the novelist Erzsébet. She is a thin, quiet woman, and even though she has written four novels, she doesn't sound self-assured at all. We talk. Her husband – a journalist and novelist, too – sits there, drinking vodka and pretending he is not interested in a discussion about women in Hungary. 'I'm lucky,' she says. 'I didn't have to work.' When I ask her what she thinks of feminism, she pauses. 'I don't understand what these women want,' she responds, glancing shyly at her husband. At this point, he just can't stand it anymore. 'You want to know who, in my opinion, was the first feminist?' he asks me, as if his argument is so strong that it will persuade me forever against feminism, his face already red from vodka and barely concealed anger. 'I'll tell you who she was – Sappho from Lesbos.' I see Erzsébet blushing, nervously playing with her glass. But she doesn't utter a word.

In a dark, smoky writer's club in Sofia, Kristina sits opposite me. She looks disappointed. Her words are bitter as she tells me about a questionnaire she sent out some time ago. 'I wrote a hundred letters, asking women if they think we need a feminist organization in Bulgaria. Everyone answered, yes, we do. But I also asked them whether they were prepared to join such an organization, and imagine, only ten out of a hundred women answered positively.' I tell her about the eight of us in Zagreb, about Jola and her group in Warsaw, about Enikö and her group of thirty students in Szeged – the first feminist group in Hungary. They were seven at the beginning. 'Ten women out of a hundred?' I say. 'But I think you're doing splendidly.' 'You think so?' she says, cheering up a bit. 'Then maybe it's worth trying.'

'*Dear B,*' I will write in my letter to the United States, '*we live surrounded by newly opened porno shops, porno magazines, peepshows, stripteases, unemployment, and galloping poverty. In the press they call Budapest "the city of love, the Bangkok of Eastern Europe." Romanian women are prostituting themselves for a single dollar in towns on the Romanian-Yugoslav border. In the midst of all this, our anti-choice nationalist governments are threatening our right to abortion and telling us to multiply, to give birth to more Poles, Hungarians, Czechs, Croats, Slovaks. We are unprepared, confused, without organization or movement yet. Perhaps we are even afraid to call ourselves feminists. Many women here see the movement as a "world without men," a world of lesbians, that they don't understand and cannot accept. And we definitely don't have answers for you. A Critical Theory approach? Maybe in ten years. In the meantime, why don't you try asking us something else?*'

14

Some Doubts about Fur Coats

Whether winter or summer, streetcar 14 never comes. They say it is because it was produced in Czechoslovakia. PRAHA, OBOROVY PODNIK, it says on a small metal plate inside. You know, a communist product – what can you expect from it? Our experience tells us that it can't stand either too much rain or too much sunshine; it runs only on nice days when the temperature is between 17 and 22 degrees Celsius and you are not in a hurry. But because it was December – the middle of winter – just before Christmas, drizzling, and incidentally I was in a hurry, number 14 was naturally nowhere in sight. Instead, in my sight was a lady, a lady in a splendid long fur coat – a silver fox, a wolf, a bear, or some other poor animal. Because of that coat, I couldn't miss her, even if I wanted to. There was a time when I'd wanted such a thing myself.

Years ago, I fell into the trap of buying a fur coat. It was winter and a cold wind was blowing from the harbor as I spotted it walking through the Church Street flea market in

Cambridge, Massachusetts. It was an old-fashioned mink coat that caught my eye from a distance, like something from a thirties Hollywood movie. I knew it was just waiting for me, for the right alibi to buy it. It didn't take me long to make one up: it was so cheap, compared to the prices in my country, where you can't buy a second-hand fur and therefore new ones are very expensive. Perhaps I can finally afford it, I thought. I put it on. In the little mirror that a young woman held up for me, I didn't look like a sick, divorced, single mother, or an East European woman at all, but like the person I wanted to be. That's what I liked so much about it: it turned me instantly into another person.

I was very well aware that for that money – I remember it cost $90, but the woman reduced it to $75 – I could buy books, something that I definitely needed much more. I also was aware that that had been my plan before I saw the coat. I was on my way to the Reading International Bookstore right there, at the corner of that very street. On the other hand, I knew that buying a fur coat for such an amount back home was absolutely out of the question; it would have cost me at least twice my monthly income. For a moment – but only a moment – I thought about the carefully composed book list in my purse, – then I looked in the mirror again and took out $80. Somewhere at the back of my mind there was one last argument left. As if someone was whispering in my ear, I heard that tiny little voice: 'What about dead animals? Do you really want to wear their furs – you, the vegetarian, animal lover, ecologist?' As artificial fur was not yet in vogue and the animals in question were long dead, I simply ignored the weak voice of conscience in my head. It just couldn't match my whole Eastern European background, which was urging: Take it, take it, here's your chance.

I have worn the coat, forgetting the whole episode. It was while I was in New York in early 1989 that I remembered it again. Another episode reminded me of it. It happened in Beograd that winter, when a young girl, J. Simović, climbed onto bus number 26. It was cold and she had on an old mink coat. At eleven o'clock in the morning the bus was almost empty, only a few old people, a housewife or two, and two young couples. 'Where did you get your mink coat, pussycat?' asked one young man. 'Did your father buy it for you?' The girl didn't answer. She stood by the door, with her back to him. Then the four of them became more aggressive. They surrounded her, one took out his lighter, lit it, and grabbed her sleeve, as if he wanted to test whether it was a real fur or not. The girl withdrew, looking helplessly around the bus. But people in the bus – even the driver – pretended not to notice what was going on. Then the bus stopped and J. Simović was kicked out in the street. 'Out!' yelled one of the girls. 'If you can afford a mink coat, you can afford a taxi!'

I read this episode in an article from the newspaper *Politika* that a friend from Beograd sent me, as an example of what is going on back home: growing poverty and frustration, leading to the revival of the egalitarian syndrome. I noticed that the reporter chose to stress that the girl was wearing an *old* fur coat. What he meant by that, using it as an argument in her defense, was that she didn't deserve what happened to her because she wasn't really rich, she was as poor as the aggressors, and therefore she was not to blame. But by saying that, he justified the same logic of equal distribution of poverty that the two young couples from the bus used in molesting her. Yet the world 'old' in that context had more than one meaning: perhaps the fur coat was inherited? Then, if it belonged to her grand-mother, it meant that she came from a prewar, wealthy bourgeois family – one more reason to be hostile toward her. But if the reporter, in his wish to protect the girl, wasn't accurate in his reporting, and she was wearing a *new* mink coat, then perhaps her father really did buy it for her. In that case, she was a daughter of a new class, the so-called, 'red bourgeoisie,' and again despised as a symbol. But in spite of a profound, decade-long indoctrination, much as it was hated, a fur coat was an object of envy at the same time. A fur coat, whether old or new, is a visible symbol of wealth and luxury everywhere – the only differ-ence is that in Yugoslavia and Eastern Europe wealth and luxury were for a long time illegal (and they become illegal every time the country sinks into a deep economic and political crisis, when, for a moment, post-revolutionary egalitarianism replaces a real solution). And even those who would like to believe in egalitarianism must ask themselves why we have to be equal only in poverty.

That same winter my friend Jasmina from Zagreb visited New York, and one afternoon we went downtown for some windowshopping in SoHo. It was cold – five degrees below zero – and we were wandering around the stores. Due to the fact that we had no money, we were enjoying the ex-pensive clothes just as someone would enjoy an exhibition of modern art. We had such a good feeling of detachment, like when you are looking at a Van Gogh painting: you love it, no matter that you know you won't be able to buy it in your lifetime. An American acquaintance advised Jasmina to go to the museums – MOMA or the Metropolitan – that she had not seen yet. 'Sure I will,' she answered, 'but we, too, have museums. It's the beautiful shops that we don't have, and I need to see them first.' We laughed at the prices, amusing ourselves by converting them into dinars. We browsed through delicate silk blouses and crazy, glittering evening dresses. We tried on Italian shoes, pretending that we were going to buy them, but they were just too little or too big or we wanted another, nonexistent color, or . . . We loved playing these 'buying games,' choosing endlessly, behaving as if we could buy if only we wanted to, discovering an element of playfulness, of fun in consumerism.

Either that day was too cold – like the day in Cambridge – or else at some point Jasmina lost her sense of humor. As we entered the Canal Jean Company on Broadway at Canal Street, she stopped in the fur department, looking at the coats there as if she was drawn to them by some magic power. 'Look, Jasmina, you don't need a coat,' I tried to convince her, as if need has anything to do with buying, especially in New York. 'No, no, I'm just looking at how unbelievably cheap they are,' she answered, hesitatingly. It was all so familiar, her argument, her wish to have one . . . She was right, of course, they were cheap second-hand fur coats, ranging from $20 to about $100. And she didn't need one, she wanted one. She looked at me seriously. 'Why shouldn't I buy such a coat? You know I could never afford

137

it back home.' By now, all the fun of our shopping spree was gone; I realized this was not a game anymore. She had on a $50 astrakhan that looked as if it was made for her. Her red lips, her pale face, her long curly brown hair looked gorgeous as she spun around in front of the mirror. 'You look lovely,' I admitted halfheartedly.

We stood there, amid racks and racks of coats – foxes, wolves, ermines, minks, real tigers, sheep – smelling of old fur, mothballs, dry-cleaning chemicals, and stagnant, un-aired cupboards. And we both knew what it was about. This was an opportunity to buy something more than a simple fur coat: it was an opportunity to buy an image, a soft, warm wrapping that will protect you from the terrible vulgar gray Varteks or Standard konfekcija coats you have been wearing all your life, an illusory ticket to your dreams. While we could easily toy with expensive dresses or shoes, the affordable fur coat was just too big a challenge. Of course, I tried the same last argument that years ago hadn't worked with me: 'Remember a poster we saw in some English newspaper, a woman dragging her bleeding fur coat? It says: "It takes thirty dumb animals to make this coat, and only one to wear it," or something like that.' Jasmina stood still for a while, then she said: 'Yes, I remember, but I don't think that you can apply First World ecological philosophy to Third World women.' With that, she bought the astrakhan coat.

Her words made me remember a TV interview by an Italian journalist with Fidel Castro, back in 1987. Surprisingly enough, they touched on ecology. Castro said he won't let his people have one car each. It is simply not possible – not only for economical but for ecological reasons. There are too many cars in the world anyway, he said, the whole of Europe is suffocating in cars. I was sitting in a small rented apartment in Perugia. It was early evening, the air was still hot, and sweat was pouring down my temples. But as Castro uttered that sentence, I shivered with cold. At that very moment, I detected for the first time in his words a frightening totalitarian idea in ecology – or better, the totalitarian use of ecology. He was asking his people to give up a better standard of living, even before they tasted it, in order to save the planet, to renounce in advance something that was glorified as the idea of progress. It seemed to me that asking for post-consumer ecological consciousness in a poor, pre-consumer society was nothing but an act of the totalitarian mind. We do live on the same planet, I thought, as his voice faded away, but not in the same world. It is precisely the Third World people who have every right to demand that the Western European and American white middle class give up *their* standard of lving and redistribute wealth so we can all survive. Otherwise, the Third World will have to pay the price for the 'development' and high standard of living of the First World in the way Castro proposed, and he definitely won't be the only one to blame.

For obvious reasons, Castro didn't mention furs in his interview, but the idea is the same, particularly because the ecological way of seeing the world is forced upon Third World people – in this case, women – in the name of 'higher goals,' a very familiar notion in the communist part of the world. And I can hardly think of anything more re- pulsive than that. So, what are women expected to do, when they see it as just an old ideological trick? Before they give up fur coats, they certainly want to have them, at least for a while; and I'm afraid that no propaganda about poor little animals will help before fur coats have become at least a choice.

This very winter I stumbled into the same situation again, this time not in New York and not with a friend, but in my own house with my mother. She had decided to buy a fur. She was saving money for it, calculating carefully where and how to buy it, and even found a shop that gives three months' credit. I felt like a character in a classic, well-rehearsed play. 'Mother,' I told her, in an over-rational, highly pedagogical tone of voice, 'you live on the coast, it's not even cold enough there.' Mother looked at me as if I were speaking a foreign language. And I was: there she was, an old lady now. Fifty years ago, she was a young beauty from a wealthy family who was to marry a factory owner's son. Then she met a partisan, Tito's warrior, 'a man from the woods,' as my grandfather used to say. Not only was he a People's Army officer, but he was hopelessly poor. But my mother married him against her parents' will, leaving her rich fiancé behind (not a bad move, in fact, because soon enough he wasn't rich anymore; his factory was nationalized by the same partisans and communists, 'men from the woods,' that my father belonged to). Her whole life she not only could not, but was afraid to buy a fur coat because of my father's true communist morals. Now he was dead. She was too old really to enjoy it. Nevertheless, she had decided to buy one. What right did I have to tell her not to? What did I know of her life, of her frustrations and renunciations

– of the clothes bought with a monthly ration card, of her desire to be a woman, not just the sexless human being propaganda was teaching her to become. It seemed to me that I just couldn't find the right words and arguments to fight her will to buy a fur coat. I felt helpless – and guilty of the same desire, too. Sitting at the kitchen table, this is what she finally said: 'You know, I have wanted to buy a fur like this for forty years.' She said that as if it was the ultimate argument, the final judgment, her last word about it.

Standing at the streetcar stop and looking at the lady dressed in foxes (I finally recognized the fur), I remembered all those doubts about fur coats. I hardly noticed two teenagers, a boy and a girl of about fifteen, standing there, holding hands, and giggling. At first they whispered, pointing at the fox lady. Then they deliberately raised their voices. 'Just imagine,' the girl said, 'that this fur coat starts bleeding somewhere near the collar. Do you think the lady would notice?' The boy chuckled. 'Oh, yes, I'd like to see that – a tiny red stream of blood dripping from each animal that was killed to make this stupid coat.' She heard them – she must have, they wanted to be heard. But her face was stone. They were not egalitarian-minded, for sure; they were too well dressed for that. Even if they were the same age as the two couples in the Beograd bus, these kids were of an entirely different breed – a new breed around here, I dare say – real ecologists. Perhaps to them and their peers their ecological consciousness is a bigger sign of prestige than a fur coat. Perhaps they feel on more equal terms with the world. I admit I saw the future in them. But they were aggressive and I didn't like it, in spite of their concern for animals. On the other hand, perhaps they are too young to understand that human beings are an endangered species and that they too have a right to protection – particularly in some parts of the world. I hope they learn this soon.

15

That Sunday, like an Empty Red Balloon

I thought it was going to be joyful and funny, the way it was in Czechoslovakia – 'a laughing revolution,' as Timothy Garton Ash calls it. This is what I saw on TV: people in Prague weeping with joy, congratulating and embracing each other on Václavské Náme'stí. Or in Bucharest. Or in Budapest, Warsaw, East Berlin. But the 1989/90 revolution in Zagreb was a cautious, sour old lady who, awakened from a half century of sleep, found herself in a land she didn't know and among people who didn't know her. Democracy in Eastern Europe has a hundred faces; this one was sad and silent.

An old plastic almond branch, forgotten in the dusty window of Building 41 after some other occasion, is the only thing that imparts a solemnly festive air. Otherwise, there is no sign that this is a polling station, except for a notice on the door of the branch office of pensioners and the Club for Older People in our tiny street, in the old part of town.

Right next to the entrance in this freshly painted yellow house with the bas-relief of the priest and historian Ivan Tkalčić, black plastic garbage bags are piled up, surrounded by eggshells, crumpled newspapers, and bread crusts. The heavy wooden door to the yard is half open. I can see laundry on a balcony above, dripping on the stone pavement, with blades of grass springing up between the stones. From the facade of the opposite building, the president of the Croatian Democratic Union (CDU), Franjo Tudjman, stares straight at us voters waiting in line, warning us that we alone should decide the future of Croatia. His poster has been pasted over an advertisement for Bibita, 'A New Drink That Cools Your Heart,' as if Europe is this sexy girl in a bathing suit, with a glass in her hand, whose secret task is to make us believe there must be a connection between her, our voting, and his photo.

Squeezed into a narrow street between Tudjman's photo and the Tkalčić bas-relief, about ten people are standing in the sun that came out only this morning, after weeks and weeks of heavy rain: pensioners who usually play cards at the club, older people in gabardine coats with worn-out collars that have seen other 'historical springs' – that of 1971 or maybe 1941. How many times have they stood in line to vote in front of houses where the postwar communist slogans, 'All to the Elections,' and 'Down with Reactionaries,' written in thick red letters, are still visible after all these years? They were forced to vote for the communists, as if the communists needed voting for. And what has changed since then? Are these elections some new trick, something that will wear out their clothing, their life itself, even further? 'I did it this morning,' says an older man, passing by and waving his hand, as if it is something that has to be gotten done with, nothing more. From the house across the street a woman comes out, drying her hands on her apron. Standing for a moment at the doorstep, confused that she is supposed to wait in line to vote, she raises her hand, smoothing her hair, as if only now has she become aware of the importance of this day. 'I put soup on the stove,' she tells her neighbor in front of me – 'and I figured I could vote in the meantime.' We let her vote; everyone understands that soup shouldn't wait, beef soup with home-made noodles. It's eleven o'clock. Her daughter is still sleeping; on Saturdays the disco clubs are open until the early-morning hours. But there is a whole day to vote. Anyhow – one should get a good rest, it's Sunday . . .

While I stand in line, I feel a strange void, almost an uneasiness, as if I am being cheated. Somebody promised me something (but who was it – the communists, the CDU president, that attractive girl Europe?), and here I am, confused, looking down the empty street on election day. All of a sudden I am not sure anymore that today is the day, one of these days that we will remember and describe in the future history books, schoolbooks, and memoirs, the day we will name streets and squares after, the date that children will learn by heart: 22 April 1990 ('Say it louder,' a teacher will tell some as yet unborn kid). But I can't help but ask myself. What will his parents remember? What will all of us remember? The street is bathed in sunlight, sleepy and quiet, completely empty. Only our small group in front of Building 41 is slowly moving forward. Through an open window nearby I can hear the sound of a vacuum cleaner, then voices from a TV cartoon. It looks like just one more Sunday morning with nothing at all happening.

Yet these people are waiting to vote at free, multiparty elections, for the first time since the end of World War II. How come it doesn't show on their – on our – faces? Why do we all look as if we are waiting in line for coffee or detergent, and why does something in the look of this place obtrusively remind me of a common grocery store? At the entrance to the election room there is a stocky comrade (something in her attitude reveals indisputably, regardless of the new manners, that she still remains just a comrade), dressed in a black dress, with a white collar that is supposed to give her a solemn, serious look. In fact, she looks like the older schoolgirl that a teacher has put at the door to keep order. Behind her, I can finally glimpse what I have been missing all morning: two flags draped over the office filing cabinet full of documents, more as if they are meant to hide it than as decorations for such a joyful occasion. Only now do I realize that there isn't one single flag in the whole street, and this is where my feeling of falseness comes from: if this is a historic moment, as they were trumpeting from different political sides, how come on this sunny sleepy Sunday morning nothing is reminiscent of a holiday? As if, deep inside, we are ashamed of doing this in the face of the communists who are still in power – as if we are traitors.

I remember my first elections, when I was eighteen: a piece of paper with names that I didn't circle, but just crossed out, so it wouldn't be valid. I didn't recognize a single name, but it was really unimportant, as everyone kept telling me. Perhaps voting didn't make any difference then, but it did look different. I can still recall the scene: a street decorated with flags, nicely dressed people sitting behind the cardboard boxes on a table covered with green cloth, and a vase with red carnations. Tito's picture on the wall was usually adorned with a wreath of flowers draped round it and two small paper flags, like the ones they gave us Pioneers at school, to greet him or the important foreign visitors sitting beside him in long black limousines.

That picture looked exactly the same as one we had in our house, the picture of Tito that my father used to decorate with ivy on May Day and hang under the window in the town of Senj where we lived at that time, while the radio played military marches the whole day. I was five then, and watching my father work on the picture, Tito seemed so huge to me: just the mole over his upper lip was bigger than my nail. Later on, in school, the teacher told us that good Pioneers visit Tito in Beograd, in his White Castle, every year. He gives them marzipan cakes, chocolate, oranges – and we must all be very good, so that perhaps we can visit him for his birthday, May 25, Youth Day. I guess all I wanted at that time was to be just that, a good Pioneer, so I could sit there and eat all that fabulous food I heard about and saw in newspapers. The photo the teacher showed the class was murky, but we could see a long table laid with a white cloth, plates and crystal bowls and flowers, and I amused myself by guessing what marzipan cakes would taste like and how oranges look, because I had not seen either of them. Pioneers wearing blue caps (with a red star on them) and a red kerchief around their necks were sitting there, some of them with mouths dirty from the chocolate. And there was Tito in his white suit, smiling reassuringly, holding a plate full of candy. A big golden ring shone on his little finger . . .

Tito's picture was hanging in this senior citizens' club, too, but it looked somehow small and wrinkled up. The wooden frame was too big for the calendar photo, cut out together with a slogan, the first line of a song we learned at school – no, in kindergarten – 'Comrade Tito, we swear to you that we won't veer from your course.' Now, we were veering away from him, and we all knew it. When I entered the room smelling of floor polish to vote, I was glad that Tito wasn't gazing at me but through a window, into the bright future that only he could see and that will forever remain a secret to us.

Later on, I was too old to vote, even to pretend to vote, giving it up because the procedure didn't make any sense at all, except for the ceremony itself. But every time I skipped it, I asked myself if *they* knew? Do they have the means to discover? Is somebody up there summing up my sins – not God, but some party bureaucrat, which in this country comes to the same thing? And does my 'bill' include even the smallest mistake? One year a tiny nervous man from the local community office rang my bell before elections (or was it a state holiday – Tito's birthday?) He was very angry. In a high-pitched voice, he explained to me that in honor of the occasion every building in the street had to put out a flag and that 'there are rules about it.' I told him I didn't know about the rules and promised I'd correct my mistake. But I never kept my promise. Was this written down somewhere, too? And now, my name isn't there on the voting list and I instinctively think of him, this little jumpy man. My disloyalty and disobedience must have come to light once, and I knew it. It surfaced now, in the indifferent voice of a young girl who, without raising her head, told me that if I wanted to vote, I had to go get a certificate from the municipal office.

In the municipal office people were waiting in a line almost to the street – the same problem: 'Not on the list.' 'What a shame,' said somebody behind me, cursing the communist bureaucracy. 'You see!' exclaimed a bearded youth, triumphant, as if there was something to rejoice about. 'Now at least you'll know who you should vote for.' Nobody answered. It was lunchtime already, but people were standing, patiently waiting for the certificate. Only this patience revealed that there was something strange going on, something more important than Sunday family lunch. If they are willing to wait, it must mean they are taking the elections seriously, I thought. But they looked too businesslike and grim. And Republic Square, the biggest square and the very heart of the city, was deserted. As I crossed it, walking toward the cathedral, the letdown seemed even bigger. From store windows and walls, torn election posters hung down in shreds, between advertisements about 20 per cent lower prices and imported cheese. What I longed for was music – any kind of music, even a brass band, people in funny paper hats dancing in the square, embracing each other, singing. Or maybe just people smiling. But that Sunday lay on Republic Square deflated, like an empty red balloon.

Suddenly I got scared by such a visible lack of joy. What are we afraid of as we circle names on a list – this time, names that we know – between two cardboard barriers in the senior citizens' club? Afterwards, we carefully turn around and look back, while folding green and pink pieces of paper, pushing them into boxes, and hurrying out of the room, as if chased by some dark shadow – uneasy and incredulous at the new democracy that came so soon, so abruptly, out of nowhere.

On my way back, as I was passing by the bus stop, I saw a young man staring at the posters with photos of candidates that had been hanging there for weeks, as if he still couldn't decide who to give his vote to. The candidates looked lost, heads turned to one side, gazing somewhere into the void beyond the posters, as if these were photos for identity cards or passports, as if all of them were taken by the same photographer, at the same place and time. It was difficult to read optimism in their serious, round, somber faces, not a trace of a smile, of self-assurance, of faith in victory. All the photos were equally gray, and the faces on them equally miserable. The more I looked at them, the more I became convinced that it's because of the poverty – the bad-quality paper, bad photos, frightened redneck faces not used to the camera, the bad food that makes the women look pudgy and the men stocky – that these elections seem so serious and pathetic, almcst like a funeral. And a funeral it was - for the past, for the well-known order of things, for the system that we hated but knew how to manage, living within it. God knows what the future will bring. One thing we knew for certain, the promises for it sounded all too familiar, as if we'd heard them before.

That Sunday, when we voted for our future in Croatia, was too silent, too normal. In the evening, an Italian journalist interviewed on TV admired the long lines of people, their patience and dignity. 'Everything was so quiet,' he said in surprise. Even the day after the elections, when the first results came out and we knew that the communists had lost, the faces of the people didn't show it. It was clear that the new democracy had its winners. But neither winners nor their followers appeared in the streets, embracing and weeping with joy and forgiving the old sins in that historic moment – as people everywhere, in all times, do when something so long expected finally happens. On Monday evening, Albanians from the edge of town held a barbecue in honor of Franjo Tudjman, the leader of the Croatian Democratic Union. They roasted a whole ox! But the big city that almost unanimously voted for him was still silent. The silence ruled, disturbing and unpleasant – as if it contained more of surprise and hesitation, caution and some vague guilt, than the excitement of victory. Perhaps because of that, the ox roasted on the outskirts of Zagreb made me feel even sadder. The cautious, sour old lady Democracy in the center of the city was asleep. A prince on a white horse was there, but it seems she didn't recognize him.

Or perhaps it was good that Democracy entered Zagreb very quietly, almost tenderly, on tiptoe, even more gently than in Berlin or Budapest, more timidly than in Prague, so silently that we had to prick up our ears to hear it and understand that, in spite of the silence, this city or this country can never be the same again.

16

My First Midnight Mass

I am the child of a Yugoslav army officer who was a communist, so everything was perfectly clear from the start: God does not exist, religion is the opium of the people, and churches are nothing more than monuments of history and culture. In our house, we never celebrated Christmas or Easter. My mother sometimes made a slightly better dinner than usual on those days, and she'd bake a cake, but the name of the holiday would never be mentioned, and we would never decorate our tree before New Year's Eve. Yet I can still remember how the entranceway of the building where we lived in Split used to start smelling of special Dalmatian pastries, walnut roll, codfish chowder, cinnamon, and ground poppy seeds. I was especially fond of the cakes that my friend Sandra's grandmother from the first floor used to make. Sandra's father was a salesman, and they were the only ones in the building who had a car and a television set – that was back in 1959, when we moved there from Zadar – and we children from the neighboring apartments would gather in their living room (the salon, as it was called) and watch Italian Christmas specials, munching almond and vanilla cookies that left sweet, sticky traces on our fingertips.

Our Christmas holiday, mine and my brother's, always seemed to come a little later, for reasons we couldn't fathom. We got our presents a week after the rest of the children in the building, and that was how we knew we were different. We never asked why; we knew it was something not to be discussed. That was what our grandmother taught us. When she took us for walks she would leave us at the church door, afraid we'd 'betray' her to our father if she let us go in with her. For Easter at her house we ate sweet braided bread, and we dyed eggs with her. 'Don't tell Daddy,' Grandma kept warning us, as if something terrible might happen if we did. Today, I am sure he must have noticed our stained hands; the dye was so hard to wash off.

My father died recently and was laid to rest with all the military honors befitting an officer and former partisan fighter. In the little coastal town where he was buried, the church bell was supposed to toll as it always did for funerals. But the bell didn't toll for him – the parish priest was away and the chaplain wasn't entirely sure whether he dared to toll for an army officer. He should have, the priest said later. The man had been born in old Yugoslavia – he was baptized and had taken communion, after all.

Before he died, my father made his peace with the Church, and in so doing beat the state to it. In 1989, for the first time in more than forty years, a public celebration of Christmas was permitted and it was proclaimed an unofficial holiday. In his own life my father had already reconciled Church and state, atheism and tradition, communism and Christianity. He was old and sick, shattered by the death of his son. Faced with that fact, all ideologies suddenly melted away, meaningless. All he cared about was that his son's burial be accompanied by a solemn and dignified ceremony.

Perhaps the thought of it reminded him of his own childhood, the scent of incense, the flickering of candles and the gilt on the altar, the sound of the organ. And so when my mother, barely able to muster the courage, asked him if my brother could be buried with a religious funeral, he acquiesced. I could hardly believe it. He, who had been remolding this daughter of devout Catholics for more than forty years, forbidding her to go to church (the wife of an officer!), or to baptize her children, or to celebrate Christmas, he sat in church on that autumn day, calm and steady. It was hard for me to look at him, not only because of his suffering but because he had been so rigid for so long, and because I felt as if I had been denied something because of him – though perhaps nothing more than a few holiday cakes – because Grandma and Mother feared him, and because we had had to lie. For a moment I felt betrayed, as if I were still a child and I was the one who had caught him out lying this time. And then, so wizened and bowed, so inwardly focused, he leaned forward and rested his forehead on his hands as if it had all become too difficult, too unreal – his son's sudden death, the prayers, the wooden pew, his own devastated life . . . I felt clearly that the armor of communist ideology that had constricted him had long ago lost its meaning. A month later he was buried by his son's side. On my father's grave a red star, on my brother's a cross.

As fate would have it, my mother went to midnight mass for the first time in forty-three years when she visited me here in Zagreb this past Christmas. She no longer had my father to keep her from it, or the fear that going would be doing something 'wrong.' 'I know he would let me go today,' she said, as if to justify herself in my eyes. Her tone suggested that she saw in my atheism something of my father's rigidity. I decided to go with her, without clearly realizing that it would be my first time ever. We went to the nearest church, St John the Baptist, a baroque structure that is more than two hundred years old, I learned from the sermon. I was familiar only with the lawn behind it, where I walk my dog. In the ten years I have been living in this part of Zagreb, I have never set foot inside the church. I don't say this with any particular sense of pride (while visiting Perugia I peered into every church, even the smallest). I simply feel that they are not something I need, though they don't bother me.

When we entered the already crowded church, I felt like an intruder as I explored the interior with cynical eyes: the blue curtain on the left that depicted the heavens with stars cut out of silver foil, the imitation marble walls, the surfeit of light blue and ocher in the frescoes, and the painting of an unknown saint on the ceiling above the first nave, from whose navel was suspended a great crystal chandelier. For me, churches are either architectural monuments or palaces of kitsch. But another image emerged in my mind: that afternoon in the television pictures from Bucharest I had seen people carrying Christmas trees home freely for the first time; in the distance I could hear the sputter of machine-gun fire.

The church was crowded – perhaps three hundred people pressed into a space built to hold half that number. Most of them were young. I recognized the faces of neighbors, several actors, a journalist, the old lady who sells flowers, the local drunk. One minute before midnight all the lights shone brightly and the church choir sang the *Kyrie eleison*. The young man in a leather jacket standing next to me sang in a deep, strong voice – out of tune, but he sang. The prayer started, then the reading from the missals. Or at least I think it went in that order; I wasn't interested in the prayers or the people standing there (although I did notice out of the corner of my eye a woman in a coat of real leopard skin leaning against the left wall, a large black hat in front of me with a peacock feather in it, and the long, refined face of a girl who sang with closed eyes). I wasn't moved by the solemnity of a rite I was hearing for the first time, but by something altogether different. I looked at my mother and saw that she, too, was singing – softly, haltingly, almost to herself, murmuring the words of the prayer in muted tones with the hundreds of people around her. After so many years, she still knew all the words! Perhaps I should have expected as much. Of course I should have. But I looked at her in surprise, with fascination, even a tinge of envy. What could I sing? I remembered a book, *Partisan Songs*, I'd been given by a teacher in second grade. My mother was mouthing the words of the church liturgy by heart, without a thought, while the only words that I could recall just then were from those combat songs. Between us there were twenty-three years in age, a war, a revolution, and another faith that had taught its children other prayers.

In the meantime, from the loudspeakers above my head came the words of the priest's sermon. 'This year,' he said, 'we are celebrating Christmas more joyously than we have in many years past.' I thought he would make only veiled references to politics, wishes for peace, 'which this silent, holy night should bring to the world'. But he was soon off in Eastern Europe, at the Berlin Wall, in Budapest and Timisoara, praying for the souls of all those who had not lived to celebrate this Yuletide holiday. 'We knew that it would happen. We have been waiting for it for decades.' His voice resonated through the loudspeakers, hovering like a prophecy over our heads. 'This is the Lord at work! The Lord created man to be free, which is why the struggle for freedom in Eastern Europe is God's work!' They couldn't possibly believe him, I thought. Why had God waited so long?

Suddenly it was all too much: Cardinal Kuharić's Christmas wishes that had been broadcast on television three times in the course of the evening, the broadcasts live from Rome, live from Bethlehem, live from Zagreb cathedral; Too much of the headlines in the newspapers, the Christmas gifts, the Christmas cards – the whole garish circus. I felt as if, after being hungry for so long for this kind of freedom, I had overeaten. If it had gone no further than the candles and the incense, the organ music and prayers, the wishes for peace . . . but the priest's baritone voice was still reverberating through the loudspeakers in an exalted account of his experience singing Christmas carols the evening before in the city's biggest basketball arena. 'Why did we have to wait forty-five years to sing Christmas carols in freedom together, in a sports hall that was not built for such things? Who was it offending?' the priest wondered with a hint of bitterness, knowing he would not receive an answer, that the answer didn't matter any more. And on he went. He spoke more, on faith, on patience – but I could no longer listen. At that moment, I was sure that my first Christmas mass was going to be the only one – at least, for me.

17

On the Quality of Wall Paint in
Eastern Europe

In the late summer of 1987, I returned to Zagreb from a two-semester visit to the States, and I almost didn't recognize my own city. It looked so different: the five- or six-story buildings on Republic Square shone, newly painted in shades of yellow, pale blue, brown, and pink, they looked like cakes ornamented with whipped cream. Perhaps for the first time I was able to notice the Art Deco sculptures on the facades, the little towers, incrustations, mermaids holding up a balcony here, braids of flowers high under the roof there, golden leaves, grapes, and chubby little stucco cherubs. Then I remembered that there used to be a warning in front of the cathedral, saying 'Danger, falling rock!' (pieces falling from its two Gothic towers), and I had thought it was just about time they did something about this city. All the same, I was surprised.

For twenty, thirty years or more – in fact, as long as I remember – nobody had bothered to do a thing. So how come the city fathers decided the city needed renewal now?

As far as I know, in my country logical reasons don't necessarily work. This decision was not made out of the blue, just because the city was dirty and needed painting and every citizen, including the city fathers, could see it. It was needed a long, long time ago – so what? It wasn't done because the citizens demanded it, either – who cares what they want or don't want, they'd better mind their own business. So, this simple painting job – or, more sophisticatedly, 'renewal' – like everything else in this state, needed a 'higher purpose'? Yes, and I soon learned that it was the Universiade, the student Olympic Games Zagreb had hosted that summer. However minor the cause, the government of the socialist Republic of Croatia, together with the city fathers, was worried about the impression Zagreb would leave 'on the world.' The world supposedly wouldn't come to such a dirty, provincial town, and so the patching up began.

However, there was one thing that bothered me seriously, and it was not the 'higher purpose' for the enormous change in the city because, like everyone else, I was quite used to that kind of 'motivation.' For me, the worst thing was the state of mind this 'higher purpose' engendered, a way of thinking that opposed Zagreb and the world so bluntly. We citizens felt that there was something offensive, even humiliating, in the way the change was done: it was done for the wrong reason and with the wrong kind of enthusiasm. If the world needs to be convinced that this city is beautiful enough to be worth visiting, then perhaps it is not worth living in. It almost looked as if the world is the bridegroom and Zagreb the bride – only a little too old and too poor, a fact that she was unsuccessfully trying to hide with her patched-up new clothes. What hurt was this distinction between 'the world' and us. I know that the city fathers didn't invent this distinction – we did.

We ourselves make it very often, and perhaps this really is the image that we have of ourselves. But we used to keep it inside, as our dirty little secret, and certainly, we wouldn't like others to think the same – foreigners especially. To see the secret discovered, to see it materialize, openly admitted, *was* humiliating in a way. If they are 'the world,' then what are we? What is dividing us; where is that invisible border? In us, in our low self-esteem, in our history and poverty, the system we live in? What makes us feel so different that we have to look up at the world, with a constant feeling that the world is looking down on us? Or is that borderline between us and them, after all, a visible, palpable one? How many times have we actually seen one, crossing the frontier from Italy or Austria, having the clear experience of plunging into something else – bumpy, poorly lighted roads, dirty toilets in the restaurants, a strange, greasy smell in hotel rooms, crumbling facades, and moss patiently climbing the walls in building entranceways, in yards, in kitchens and even in bathrooms. As if all these things were already forgotten, even if we still live with them. As if they didn't belong to us really – as if we are just in transition. Or as if we gave up the fight against time and nature, against the price of wall paint and its quality – all in advance.

Perhaps it is color – paint – that matters, after all. Today, after less than three years, all the newly painted facades, the nicely decorated pastel 'cakes' we were so proud of, are old again, dirty yellow, brownish, dirty gray, like in old photos. They all look dusty, covered with a thin sepia film that seems to stick to shop windows, lamps, door handles, streetcars, neon signs, benches, even trees. This film is like a fingerprint of time, its personal signature. We recognize its power and its strange effect on us – we give up. If I had seen it only in Yugoslavia, this phenomenon of facades darkening so quickly, I would think it was an endemic disease or ascribe it to a psychological effect: entering the country, one immediately becomes pessimistic, capable only of seeing everything in slightly darker colors. But I saw it in Prague, too – 'renewed' Prague, that is – wandering among the small streets of Stare Mjesto, Pařižská, Kaprová, Š'iroká, Dlouhá. The unpainted facades looked even dirtier because of the newly painted ones. And the painted ones were slowly but visibly absorbing the sepia color from the very air, from the breath of the people who passed by.

And I've seen it in Budapest. I arrived at midnight. From the railway station, I walked down Rákóczi U'tza, then József Körút Boulevard toward the Nemzeti Hotel. The boulevard, the shopping center of the city, was not only empty but almost dark. Or so it seemed to me, because I experienced the 'light shock' that usually happens when one enters any Eastern European city from the West. The contrast is such that for a moment it has to cross your mind that either the city has had a power failure or there is danger of an attack from the air. The street lamps in Budapest were casting a scant yellowish light, and besides, every second one didn't work at all. The shop windows were not lit, most of the neon signs were off, no streetcars passed by at that time, only a car or two – a ghost city. In a moment of inspiration, an American colleague described a similar view of Prague: 'The scattered lights of the town are warmly orange, and their glow is subdued enough so that the stars and a swirl of moonlit cloud are crisp and sharp in the sky.' But where he saw romantic 'scattered lights' that made him look into the sky, I saw nothing of the kind. Living in a country where life functions in pretty much the same way as in Prague (or Budapest), all I could see was the saving of expensive electric energy, bad bulbs that burn out too quickly, broken lamps that take years to replace. I admit this might give a romantic impression of the city, especially if you happen to be American, but it is surely unintentional – has anyone ever heard of a romantic communist regime?

The next morning, from the 130-year-old Café Gerbeaud on Vörösmarthy Square, Budapest did look different. It was a sunny day, 'crisp and sharp,' as my colleague would say, yet I couldn't get rid of the feeling that the square, the beautiful buildings on Castle Hill or Belváros, all six bridges over the Danube, the very city itself was covered with the same sepia-colored film as Zagreb, Prague, or East Berlin – as if it were fading away, crumbling at the edges, disappearing. Entire pieces of facade fell down, parts of walls, steps. I saw statues of giants without noses, angels without wings, caryatids without breasts, horses with broken legs, forming the imaginary menagerie of a long-forgotten world . . . The city is slowly decaying and nothing, not even the new government, can stop this irrevocable process. The West German poet Hans Magnus Enzensberger, in his book of essays *Europe, Europe*, maintains that 'It is this ubiquitous erosion that constitutes the secret, the terror, and the charm of the metropolis on the Danube. An irresistible, spontaneous, higher force: entropy. The painter who dips his brush in the pot knows that his work is futile and that only one thing can be relied on: really existing time, which takes hold of and conserves everything, even as it wears it down. History is a process of erosion. What we call socialism is only its viceroy.'

There, in Budapest, I realized where that inner energy comes from, the energy that is eating up every city in Eastern Europe, its Gothic or Baroque or fin-de-siècle crust, its 'gemütlich' skin, its nostalgic beauty, making shabbiness and the color of sepia their common denominator. To say it's the poor quality of the paint under socialism is correct, but it is not enough. To say it's soft-coal exploitation and air pollution, bad gasoline and bad cars, or lack of money – that again would be correct. But not the whole story. All these reasons (and probably many more) are not enough to explain the decrepitude. I think the reason is in us. The cities have been killed by our decades of indifference, by our conviction that somebody else – the government, the party, those 'above' – is in charge of it. Not us. How can it be us, if we are not in charge of our own lives?

I live on the ground floor, and my room looks out onto a busy street in the center of town. I consider it an advantage to live on this level because I see things I otherwise wouldn't notice at all. I watch children, dogs, and drunks. I see angry people discussing politics in front of my window (so I knew which party would win the first free elections in my republic) and housewives telling each other what there is new to buy (or not to buy in the grocery store at the corner. Every morning, even before I get up, I get new information. Besides that, what I see the most is litter: I can carefully observe how people rudely, carelessly, mindlessly throw away newspapers and plastic bags, cigarette boxes, bottles, umbrellas, old shoes, even old refrigerators. They toss them out, slap them down, kick them – using the whole pavement, the flowerbeds, and the street as a huge garbage can, assuming that that famous 'someone' will take care of it again.

Every public space is like a billboard, with messages from the collective subconscious of the nation. There one can read passivity, rage, indifference, fear, double standards, subversion, bad economy, a twisted definition of 'public' itself, the whole *Weltanschauung* – an entire range of emotions and attitudes accumulated and exposed. We behave as if the public place belonged to nobody. Or, even worse than that, as if it belonged to the enemy and our sacred duty is to fight this enemy on his own terrain, perhaps even to exhaust him. Public space begins outside the apartment. But the problem is that in our mind, public equals state equals the enemy. If you can't destroy the system, you can certainly destroy a telephone booth, ticket machine, parking meter, or flowers in a park.

In this silent war, the losers are our cities. Entranceways into buildings not only serve for entry but double as toilets, as well as places to express your deep inner feelings. These feelings (if you can manage to ignore the smell) are hatred and destruction. Perhaps there are not always messages scrawled on the walls of the corridors, because this is the 'better' part of the city (though even here locked doors don't last long), but there are more subtle traces of the frustrations: broken glass, unscrewed light bulbs, garbage corners, demolished fences, an elevator that doesn't work, doorbells ripped out. Some people use parks for recreation, for enjoyment. Others adapt them to their own needs. That is, to express their rage: they destroy benches, twist iron fences, shatter trees, make abstract figures out of garbage cans – or use the park as a toilet, too. It simply is not their place and there is no money in the world that can restore the damage, only a change of attitude. When I first went to Hyde Park in London, what surprised me most was not Speaker's Corner – I had read about that – but the people sitting on the grass, sunning themselves, eating, or even sleeping. In our parks, I've never seen that. Grass is for looking at, not for enjoying.

Maybe only now, after the political changes in Eastern Europe, will we have a chance to repossess our cities, re-privatize them, treat them as if they are not merely places we are sentenced to be in or which we only pass through. But to do that, we first have to realize the meaning of a public space in general – not just facades, but lights, pavements, garbage, toilets, parks, and house entrances. We have to understand that the public space is not a Kleenex that one uses, crumples, and then throws away. It's more like an old-fashioned, fine batiste handkerchief, embroidered at the edges, that one has to wash and iron to be able to use it again . . .

György Konrád wrote that trains and movies run slower

here, because time runs slower. I think it runs differently. As layers and layers of illusions are peeled away – the illusion of beauty, the illusion of power, the illusion of importance, even the illusion of meaning – time profoundly changes our view of life itself. The Austro-Hungarian Empire built up its signs of wealth and power for four hundred years. They slowly decayed, fading away. Then for almost half a century the communists tried to destroy the past and replace it with their own symbols – they faded even more quickly. Now the new governments are again changing the names of streets and squares, destroying old monuments and replacing them quickly with new ones, taking history and memory as their own little playground. But the cities are remembering and showing it, and the people, too. The nostalgia and hopelessness of the Central European soul, its sadness and cynicism – the inner sepia, if you wish – all stems from this. So, I guess, we are something else, after all, something visibly different. In our cities, 'renewal' does not renew, it only points out the passing of time, the fact that there is no progress, that history repeats itself endlessly.

The houses are like the faces of people, that is what I like best about them. Their beauty is not only at the surface. This is why cities become mirrors of our inner selves. Their genuine beauty is in their perfectly human size, in the feeling that you can master them, not that they are mastering you. They have a center, an orientation point, a main square, main streets and buildings. And people need centeredness: it gives them security. A city without an old town is like a shell without a pearl. That is why I think that the postwar reconstruction of Warsaw's Old Town or Budapest Castle makes perfect sense – without it, a European city would be as if without a core. As for 'the world' – assuming this division, outside as well as inside, already exists, let it come here. Perhaps this is the only place where it can see that history in essence is a metaphysical category.

18

The Day When They Say That
War Will Begin

That day, as on every other, a truck with bread stopped near my house, and I watched two men carrying baskets full of long, golden loaves, round *pogače*, brown rolls, *pereci*, kaiser rolls, pitta bread, and croissants into a nearby bakery shop. As I opened the window to let in the crisp, blue winter day, the smell of warm bread entered my room. I took a deep breath. Everything looked so peaceful, so normal: the movements of the men carrying their baskets, people passing by, children's voices in the playground . . . for a moment I even thought I heard a bird humming – a harbinger of distant spring.

I went to the kitchen and fixed myself a cup of coffee. And while the bitter, dark brown liquid slid down my throat, I switched on CNN. The smell of bread, the voices outside, the peace and quiet of a normal life all disappeared, as if the new day had been sucked in by a whirlwind from that blinking screen. Instantly, I found myself in the middle of a battleground: it was the eighth day of war in the Gulf,

Thursday, 24 January 1991. But it was also the third day after the end of our federal presidency's ultimatum to surrender all illegal arms – the day when they say that war will begin in my country. Sucking our reality in, the TV screen chewed it up and spat out a new reality, only without us. On CNN, they were speaking only of that distant desert war, with its tons and tons of bombs, of helicopters and ships, as if our conflict is too small for the world to take notice of. After all, it's only some 20 million people, lost in the hills of the Balkans, with no world powers involved and no oil fields to make it important.

Indeed, if it should happen, our war would be on a small scale, domestic – almost like a domestic animal, but no less bloody because of that. Our people know each other's weak spots, they know where to stab so it hurts the most. These wouldn't be enemies with unknown faces, speaking strange languages – here, Serbs and Croats from the same country, the same town, maybe even the same streets would fight one another, understanding each other perfectly, continuing where they stopped in 1945. It will be their private little war, and they know it. In a way, a military coup that caused a civil war between Serbs and Croats would be only a continuation of the conflict between and during the two world wars. The memory of it is still alive; many politicians now in power fought in the war. With more than a million war victims, almost every family lost someone. And they were not all killed by Germans or Italians. Many times the enemy was a nationalist extremist, a Serbian *četnik* or a Croat *ustaša*, and it seems we still have to deal with this fact, because we haven't been able to let the dead be dead. During that war,

Croatia was a fascist puppet state run by the *ustaše* under German protection who built concentration camps for Serbs, communists, Jews . . . There still is an unresolved dispute going on about one such camp, Jasenovac: Serbian and Croatian historians, can't agree how many victims died there. For one side, the number is too small; for the other, it is too big. Besides, stories about the dead and disappeared that I keep hearing from my family alone make me feel as if the war has never really ended.

For forty-five years, within the iron embrace of the Communist Party, the wounds of nationalism were not healed. Instead, they were ordered to disappear. Nationalist antagonisms were suppressed and replaced with 'brother-hood-unity' ideology. As kids, we all read in textbooks that brotherhood and unity were the greatest achievements of the communist revolution in Yugoslavia, learning by heart *bro-ther-hood-u-ni-ty, bro-ther-hood-u-ni-ty*, and shouting it at mass meetings on May 1 or Republic Day, on squares and streets, clapping our hands and repeating endlessly these words, as if they represented some magic formula for our survival. We children didn't understand it; our elders perhaps didn't believe in it; nevertheless, we were supposed to stand up for it. In the ten years since Tito's death nationalism has started boiling in this country, like a steam kettle, which is now whistling loudly, becoming our only reality.

After the first free elections in May 1990, when new, non-communist governments were elected in both Croatia and Slovenia, the country split in two. Or almost: there is such a big gap between the Serbian idea of the federal, centralized state, under its control, and the Slovene and

Croatian idea of a loose confederation, giving the six Yugoslav republics the possibility of seceding and becoming sovereign states. The problem is that Serbia has the Yugoslav People's Army on its side, as well as the collective federal presidency under its control. On January 9 the presidency, under army pressure, issued an ultimatum to the so-called 'illegal paramilitary troops' (meaning the Croatian police reserve) to surrender arms in ten days. They didn't. Then they were given two more days. Again, the arms were not turned in. Now the army warned that it would start to collect these arms, searching houses, one by one. Everyone knew it could mean only one thing: a clash between the army and the Croatian police, civil war. And January 24 was the third day after the ultimatum expired.

When the political changes swept through Eastern Europe, I didn't imagine it like this, so close to the possibility of war that it makes your blood thicken with fear – a fear that came slowly, gently, creeping into our lives, so that we didn't even recognize it at first. The police troops in front of the parliament and other important state buildings, the TV and radio station, posted there at the beginning of the ultimatum, didn't seem to worry anyone. We still didn't connect them directly to our lives. There was no curfew, people walked the streets in crowds, shops were full of imported goods – French cheese and wine, Norwegian salmon and caviar, American cigarettes. Besides, Croatia is not Lithuania; we are not occupied by Soviets. Enemies, war, Serbs, the army, danger – it all existed on another level, on the abstract level of daily news, of political speeches, of pictures – like the distant sound of thunder, background music, a noise that you could dismiss with a little effort, thinking that this must pass. Yet – rumor came closer and closer, it surrounded us from all directions, until we finally realized we were right in the middle of it: special police keeping watch night and day just two houses down the street, an emergency session of parliament being broadcast nonstop on TV and radio, empty streets in the evening, people talking about whether they should leave the country, lights in the houses that went out later and later every night – until Thursday the twenty-fourth arrived, with its smell of fresh bread.

What should a person do on that day? Just sit and do nothing? Still thinking no, no, it won't happen, not here, not now, clinging to the kitchen sink as if that cold, solid metal is the sole definite point in reality, and if you grab it tightly enough, it will hold your world so it will not fall apart that very instant? It's morning, you are alone in your ghostly house, and you don't know what one is supposed to do now on this strangely quiet-looking day.

I did what I do every day. I walked to the grocery store on the corner to buy something for lunch. 'Lunch on such a day?' I could hear my own inner voice arguing with me. 'Yes, lunch, because all this is surreal, it's absurd, war doesn't come like that!' I also could hear my answer, as if I knew – from stories, from literature, from movies – how war begins. In front of the store I noticed a strange activity. People were parking cars and filling them with cardboard boxes full of flour, sugar, oil, rice, pasta. In this country, that could only be the sign of prices going up enormously at once – say 300 per cent – or else a warning that there is an even bigger danger in sight. Shelves in the shop were half-empty, as if people were buying anything they could stock up on for the hard times to come: toilet paper, shampoo, detergent, tomato puree, canned beans, cheese. A woman standing in line at the cash register in front of me bought twenty liters of oil. 'We sold out all the flour and sugar in a couple of hours this morning,' said the shopkeeper. 'We made more money than at the Christmas holiday.'

I bought the usual stuff: fresh bread, yogurt, oranges. I just didn't want to take part in this hysteria, becoming a fish caught in a net of war that someone was pulling, tighter and tighter, around me. For the moment, I thought I could escape it by just going on pretending: if I behave normally, everything will be normal. But it seems that everyone else was caught in it already. My friend told me that her grandma ordered her to buy twenty-five kilos of salt and was issuing instructions for how to keep flour in a loft: it has to be kept in a dry place with enough air, in boxes, sifted from time to time. While I could understand why someone would buy flour, I just couldn't understand twenty-five kilos of salt. 'Grandma says it's because you can sell it later on, or trade it for food, it serves like money,' my friend answered. 'And she also says that with a bag of pepper, using it the same way – especially if you trade it with people from villages – you can pass through the whole war without being hungry. But I persuaded her not to spend money on that. 'Pepper is expensive and the war can't last for years, can it?' Twenty-five kilos of salt? A bag of pepper? It sounds crazy, but what if ... Then my neighbor Andrea passed by. She bought her staples, too: chocolate bars, a bottle of vodka, and cookies. 'I'll just sit in front of the TV tonight and wait for the military coup. I only hope that by the time it starts, I'm dead drunk,' she said. She was standing in my doorway, all her defenses against the war contained in a single paper bag.

When she left, I opened my kitchen closets and found myself staring at my one pound reserve of salt (because, of course, salted food is unhealthy), no sugar and no flour (because it makes you fat), a little vegetable oil. But I had some dog food, muesli, dried fruit, nuts, miso, brown rice. I calculated that my little family – my daughter, our two dogs, and me – could survive three days straight. Or perhaps even four, because she is always dieting. I told her that we too should get some staples, but, eating her portion of bran and cornflakes, she laughed back at me. 'Flour, sugar, oil? Who needs it. It only makes you fat.' Then she left for school. I asked her not to go, not today. 'Don't be paranoid!' she answered rudely, hiding her fear with hard-boiled cynicism – her fear of the unknown and of a horrible something that she felt was approaching and that she didn't want to show in front of me. I knew it. We used to play this game of hiding fear from each other every time there was a dangerous situation, imagining this would make it easier for the other. We know that we play it. We know that it doesn't really help. As I stood there, watching her applying a bright red lipstick (lipstick!), then putting her coat on, I became almost physically aware of her going away, as if she were still a baby, a part of me, of my own body. *There is no way to protect her from this madness, from the dark molasses of war that is clotting around us, gluing us all together in an immense mass of people where no one can be distinguished anymore. There is no us — me or her — anymore, I thought, giving her a light kiss on the cheek.* For a while, the smell of her heavy black hair fluttered in the air after her.

Now, alone in the kitchen, I listen to the radio; the newscaster is announcing that today the army commander has ordered the highest state of alert, soldiers have been given dried food rations for forty-eight hours. The president is asking the citizens of Croatia not to do anything that might provoke army reaction. 'But if it comes to that, if the army attacks, then we will know how to defend ourselves,' he says. As the reporter talks about last night's army maneuvers, about people hearing tanks on the streets, I feel as if his words are climbing up my throat and strangling me, as if they are hands. *Too late*, I think, sitting alone and listening to some waltz that has replaced his words, sliding on the floor and embracing me with the sweet, comforting thought that it is not true, just a voice on the radio, nothing more . . .

How to recognize the beginning, how to know that the war is here? Should I wait and judge only by outside signs, listening to remote gunfire or a soldier's footsteps in front of my door? Is it only then that I will realize that my doors are so weak, made of a thin wooden frame with glass, that they couldn't protect me at all? While I watch the central TV news broadcast, my house and the nearby streets blanketed in a milky winter fog (some kind of protection, I think), I realize that by now the fear has gained autonomy, and it has become completely unimportant to know where it stems from – the knowledge of the source wouldn't help me now. Fear is like a beast that gnaws at you, eating you up bit by bit, until you totally surrender to its teeth, and you don't even think that there might still be a chance. A chance of what? Of taking a passport and some money and leaving all this behind? But to go where? And when? Now – or when I can actually see tanks rolling down my street (does it have to be *my* street for me to believe it)? But to decide, I need to think of the future – something I can't think of, not anymore. For the moment, when I think of the word 'future,' I see it as a train from Claude Lanzman's movie *Shoah*: its destination is known and yet it is not possible to step down from it. Not only because all the doors on this train taking us to our final destination are closed – they are, but there is always a way to escape – but because of hope, a hope that brings us death as well as keeping us alive, a hope that makes us sit still, waiting, waiting . . .

As the soft, misty darkness starts to fall, I can feel something cold, like a piece of ice, growing inside me, spreading in my chest, drying my mouth, making my palms sweat, making my body shake with illness yet unknown. In a certain moment, a person stops thinking politically, stops thinking at all. The fear finally takes over. You can tell it by the way you stay there, in the middle of a room that you don't recognize anymore, staring into emptiness, paralyzed by a sudden numbness inside you. No thoughts, no movements, nothing but this crystal moment of pure fear shining inside you. It's not the fear of death but of planned death, death invented in someone's head, death as a statistical number, a mass death in a deadly game of power. It's the consciousness of how fragile and transparent the line between life and death is and the knowledge that you are about to face it . . . Then the tiredness of this day, of waiting for the war to begin, engulfs me. I imagine this is how people feel when they are about to freeze: the white, soundless night when death takes you gently in its arms, closing your eyes and putting you to sleep. There is no pain, only heaviness in your limbs, in your lungs, in your heart, beating slower and slower.

But in that moment, when fear finally surges to the surface howling like a beast through my throat, I remember candles. *I have to buy candles, I have to buy candles* - I repeat to myself. *Sure, they'll cut the electricity, there won't be light, how can we live without light?* I hurried to the store and bought a box of candles and a big package of dog food. It made me feel better – at least I am *doing* something. Walking home from the store, I saw a group of kids coming out of school. One girl had on a pink parka, and its innocent color hit me, as though someone had slapped my face: but perhaps it will not happen . . .

As if relieved of some heavy burden that was pressing me to the ground, I started cooking dinner. Soon my daughter will come in, and this is what I should do – cook dinner – because, suddenly, I am not afraid anymore. The war is here. I recognize it now. It tricked me – it tricked all of us. It's in our waiting for it to begin.

19

How We Survived Communism

Vesna held up the pantyhose to the light, put one hand in for inspection, spread her fingers, and then slowly, looking for a run, pulled the hand out. One leg was good. On the other she discovered a run. 'This pair is no good,' she said, putting it aside and reaching for a new one from the pile in front of her. We were sitting in her kitchen on a bright Sunday morning. I had come across to borrow a vacuum cleaner, as that very morning mine had decided to give up its long, fruitful life. Sitting across from us, her mother reached for the rejected pair. 'But, Vesna, what a shame, these pantyhose are still good to wear around the house. I'll take them to a repair lady to be darned.' Vesna looked at her angrily, then burst out laughing. 'No, you won't,' she said, and turned to me. 'Every time I try to throw away a pair, she tries to "save" it, to repair it, to fill a cushion with it, tie a garbage bag or filter home-made juice – she can find tens of different uses for old stockings or pantyhose. It's like she still lives in the fifties.' Her mother calmly sipped her coffee, shaking her head as if we couldn't understand her however hard we tried. 'I don't live in the fifties, but you never know what might happen. Don't you remember, just a couple of years ago there were no panty-hose to buy. Then you asked me to darn the old, "saved" ones for you. I sewed them with a nylon thread over a light bulb (thank God I saved the old bulbs – I'll never stop re-gretting a good old wooden mushroom for repairing stock-ings that I threw away). Besides, to throw it out just because of one run . . .'

I said I didn't know there were any repair ladies left – after stockings had become mass produced, though not cheap. Not in these parts. She said there was one, just one left. She closed her shop, and she works at home now. I could see her at her table with a lamp and a funny little stitching machine. Perhaps it is the same lady on Maksimir-ska ulica I used to visit once a month with a bag full of stockings to repair. 'Will you wait?' she would ask me, knowing that I had already prepared one pair, with 'just one run,' to take back home. It must have been more than twenty years ago; stockings were rare and expensive items then, and only after many repairs would one throw them away. 'Look now,' Vesna's mother continued, 'what do you know, a civil war might break out any minute: Serbs would fight with Croats, Czechs would fight with Slovaks, Hun-garians would fight with Jews. How can you be sure of any-thing?' 'But, Mother, if this happens, then it will such big trouble that nobody will think about a shortage of panty-hose,' protested Vesna. 'You'd be surprised, my dear, to know that people have to live and survive during wars, too.

Besides, how do you think we survived communism?'

Yes – how did we? Certainly not by throwing away useful things. Generally speaking, in any communist country there are not many things to throw away. One could even say that a communist household is almost the perfect example of an ecological unit, except that its ecology has a completely different origin: it doesn't stem from a concern for nature, but from a specific kind of fear for the future. Such an ecological unit – like any other – has two basic principles, collecting and recycling. You recycle, recycle, and recycle, redefining an object (pantyhose, for example) by turning it into something else, giving it one function after another, and you throw it away only when you have made absolutely sure (by experiment, of course) that it can't be used anymore. But in order to recycle properly, ie usefully, you first have to know what to collect. Collecting principles, so to speak, depend greatly on different kinds of experiences in different communist countries, or – better still – on different *degrees* of poverty. But they basically can be divided into several categories: *general objects* (old cloth, shoes, household appliances and furniture, kitchen pots, baskets, brooms, newspapers); *objects that normal people in normal countries usually throw away* (otherwise known as packaging – bottles, jars, cups, cans, stoppers and corks, rubber bands, plastic bags, gift wrappings, cardboard boxes); *foreign objects* (anything from a foreign country, from a pencil or notebook to a dress, from chewing gum to a candy wrapper); and *objects that might disappear* (a very broad and varying category, from flour, coffee, and eggs to detergent, soap, pantyhose, screws, nails, rope, wire, perfumes, notepaper, or books – you simply never know, you never can predict what will be next, which, after all, is the primary reason for collecting). And while I am perfectly aware that poor people in not so poor (Western) countries collect and recycle according maybe to the same principles, people in Eastern Europe were, in the first place, almost all poor enough to have to do it. The other main reason is that they live in a state of constant shortages, never sure what they will find in the shops next day.

I washed the floor in my house with an old pair of man's pants, never realizing how odd a broom dressed in pants looks, until my friend, a foreigner of course, pointed it out to me, laughingly. But, just like Vesna's mother, I thought it was a pity to throw them away, when they could be re-named and reused as a floor mop. Couldn't you buy a rag mop, my friend asked. Yes, but why? My grandma did the same, my mother, too – besides, until the late sixties there were no floor mops to be bought. Long experience proved that genuine cotton underwear has a fantastic ability to absorb dust and wash floors, windows, tiles, and so on, that is why my mother still uses it.

Then I realized that this friend of mind didn't know any-thing about recycled clothes: reknitted pullovers, old coats turned inside out and made into children's coats, or a new sheet made of two old ones. She wasn't aware that in the 'ecology of poverty,' nothing is wasted – especially clothes. They're not usually given away ('Am I worse than you?'), ex-cept to the Gypsies. A person might get offended, since new clothes are proof that you are better off than the others. We don't only use them – that's the second stage – we wear them first as something called 'around-the-house' clothes. An unaccustomed person, coming into the average household, could see the strange yet usual sight of an otherwise respectable, even important person – a university professor, let's say – dressed in striped pyjama trousers, an old pullover eaten by moths or mended with wool of another color, slippers, and a worn-out bathrobe. And because this kind of 'collecting' is a national sport, nobody minds being seen in these absurd rags.

The living conditions kill all privacy – or spread it out to the whole community, if you wish. Apartments are too small, too crowded, or too divided, and either way you are bound to meet other people on your way to the kitchen or the bathroom. Since there is no such thing as a self-sufficient communist household, you depend fatally on your neighbor for all kinds of favors, from borrowing coffee or sugar to washing, cleaning, or cursing politics – or getting your child enrolled in a better school. He or she will inevitably see you in your 'around-the-house' clothes. Perhaps there is a good side to that; people don't have any illusions about each other.

While there is some obvious logic in collecting and recycling old clothes, it's hard to find any logic at all in collecting objects that are meant to be thrown away. That is – if you don't live anywhere near a communist country. For example, why would somebody keep an old shoe box? Once you bring your new shoes home, you simply throw it away. But a nice, strong shoe box can have several purposes. The two most usual were – and are – storing photos and storing old bills. People keep them in the bottom of a closet or at the top of a cupboard in a bedroom. When you ask to see the family pictures, they don't pull out a photo album, but a dusty shoe box. They untie it carefully in front of you, as if that single shoe box holds something very precious, a piece of their lives, something that shouldn't be looked at every day. The reason for the shoe box is a simple one: for a long time there were no albums to buy; then they were very expensive. Besides, it seems that people here don't really have so many photos, and they don't look so often at them or show them around. Shoe boxes are fine for storing bills, too, particularly old ones - very old, in fact, ten, twenty, or more years. Utility bills, rent or credit bills . . . when you are dealing with such a vast and inefficient bureaucracy, you have to be prepared to go back years and years to prove your innocence. A shoe box is almost like a computer, full of neatly stored data necessary to survive in a system that is designed to destroy the individual.

But in every household the absolute priority belongs to collecting glass jars, and I think it is because you can store other objects in them that you collect. Again, what could be more normal than to throw the jar away after eating your pickles or jam – ideally, into a container for recycling. But no – because, living here, you'll soon find that there is another way of recycling it. There's always something that needs to be stored in it. A handful of rusted nails, or number 1 screws (you know that when you look for them in stores, you can never find them), or perhaps buttons or rubber bands (but you have to take those off packages first), or plastic stoppers or corks, old razor blades (you never know when these will disappear from the market), pieces of soap or string (just in case), breadcrumbs, coffee beans, fresh garlic – or maybe even home-made pickles. In bigger jars you can store some food and then keep it on a balcony during winter, like in the refrigerator – a very handy invention. What's more, if a jar gets broken its cap can be turned into an ashtray. With virtually everybody smoking, who has enough ashtrays, anyway?

Most people collect cans, especially big ones – but small ones will do, too. 'It's a shame to throw this away,' my neighbor, an old lady, would say, and she planted red geraniums in dozens of cans she would put on her windows, in halls, on steps, in the bathroom. I think women actually prefer 'gold' to 'silver' cans; they do look nicer on the window sills and balconies, in small yards behind the house, even in the house – wherever you expect to see flowers. You take a can, a nail, and a hammer, turn the can upside down, make a hole with the nail and hammer, and there you are – a new vase, practically out of nothing. Now, that's what I call ecology. A flower can grow in an old washbowl or a pot, most probably sky blue or bright red with white dots. Plants stunted from lack of sunshine in cities look a little bit more lively like that. That is, if such a luxurious use of the old pots is permitted, because they usually end up under the sink, filled with potatoes and onions. Perhaps they see the light once or twice in ten years, when a kitchen or a room is painted (in these parts people still do it themselves) and they serve as a paint pot.

'Even today,' says Vesna, 'I can't get rid of the habit of washing plastic yogurt cups. In the mid-sixties when they started producing them (before that, yogurt was sold in small glass jars that you had to return), I was a school kid and we used them for watercolors. At home, we girls played kitchen with them, or drank out of them, or kept salt and sugar in them. Today, even if I don't use them, I just collect them, God knows why. I guess nowadays collecting doesn't reflect the state of facts as much as the state of our minds. We are hungry for things and afraid of the future – it's deeper than I thought.' Her words made me remember another kind of collecting, another kind of hunger – my hunger as a child for a nice cellophane candy or chocolate wrapper that I could get only from a friend at school whose father lived abroad. I would press them between the pages of a book, and then look at them, at the foreign words like *framboise* or *sucre* or *chocolatier*, still smelling of their extravagant, delicious contents I had never tasted. While 'abroad' (at that time 'abroad' was a category that included everything beyond the border, we made no distinctions) they wrapped each candy in a beautiful paper, the only kind of candy we had was called *505 sa crtom*, and it came in a red metal box, only as a present for New Year's (not Christmas, mind, Christmas didn't exist officially). No wonder, because at that time, in the early fifties, there was only one candy factory, named after the war hero Josip Kras.

Of course we were fascinated with these little wrapping papers – our first contact with something foreign – sensing that there was something still more unknown and desirable out there. Later, we collected foreign tags and stickers of any kind – from candy, beer, cheese, clothing, – then foreign cigarette boxes, beer cans, or Coke bottles that tourists would throw in the sea while waiting for the ferry.

The only important thing was that they were foreign. Why? Because everything foreign, from wrapping paper to a beer can, was more beautifully designed and, surrounded by poverty, we were attracted to this other, obviously different world. Much, much later – perhaps too late – we learned that it was all because of consumerism, just to attract buyers. But I don't think it convinced us, because even today we are passionate collectors of foreign objects, as if we are still trying to possess this mythical 'abroad' or the imitation of it.

It's a perfectly familiar feeling: I can see my five-year-old self going into the bathroom, reaching for my mother's lipstick, taking my chewing gum out of my mouth, putting some lipstick on it, and mixing the chewing gum until it became pink; then going out and pretending to the girls it was the original 'Bazooka Joe' (even if I tasted it only once in my life, and only one piece), so they would envy me, praying that some of them don't ask me to show the wrapper with the little comic strip (the first comic I had ever seen) inside, with figures talking in clouds in some unknown language from the moon.

My grandma died in the seventies, but before that she spent a couple of months in the hospital. My mother took this opportunity to clean up, at least a little, her cupboards and drawers of 'trash,' as she called it, because my grandma was famous for collecting everything in sight. The contents of the massive old oak cupboard were rather normal. Only old coats (I particularly loved a green one, sewed sometime between the two wars, with a silk label: *Modewerkstatte Franziska Bundschuh*), a balding astrakhan fur stinking of mothballs, several pairs of almost new hand-made shoes, gloves, a pile of sheets, my mother's baby dress from her baptism. Another cupboard was, in fact, a small warehouse or a boat navigating unknown waters. It was full of neatly stored detergent that had turned almost to stone, bottles of rancid oil, several kilos of sugar, flour, and coffee (apparently the household staples), some packages of tea, biscuits, pasta, cans of tomato paste (she loved Italian cooking), beans, and even a kilo or two of salt, in spite of the fact that nobody remembers a shortage of *that*. The food was stored on the lower shelves. On the upper ones was everything else, such as a roll of white tulle, quite a bit of wool in different colors, brand new and repaired pantyhose and stockings (I believe even from before World War II), black and brown hair dye, shampoos, soaps, hand creams, toilet paper, out-dated antibiotics, aspirins, insulin (even though nobody in the family is diabetic) and some other pills without labels, absorbent cotton, and about five or six packages of sanitary napkins. Rather than a warehouse, her cupboard looked like a museum of communist shortages.

We were not surprised – not until we opened one of her drawers. That was too much, even for us, collectors our-selves. The drawer was full of plastic bags. Washed, dried, and sorted, then tied into bundles with rubber bands, there were bags in all sizes and colors. Large ones, from foreign department stores or famous shops that we had brought back from our trips to Austria, Germany, Italy, Spain, or Sweden maybe twenty years ago; then smaller ones from shops in Zagreb; then the usual ones, without labels – down to the smallest ones. Like an archeologist, her collection documented the development and use of plastic bags ever since they came into use in this country, with the rise in the standard of living (when they were given out for free) through the economic crisis (when they disappeared) to the present time, when one has to pay extra for them. Friends returning from the USSR, told us that plastic bags are much in demand, especially with labels, because women there carry them as handbags. We didn't, but we washed them, finding new ways to use them, until they ended up in the garbage – as garbage bags, of course.

I think these drawers of my grandma's show not only how we survived communism, but why communism failed: it failed because of distrust, because of a fear for the future. True, people did collect out of poverty, but a very specific kind of poverty, a poverty in which the whole country is deprived, everybody is poor, a poverty when to be poor and deprived is a state of life that hardly ever changes, because it cannot be changed by words, declarations, promises, or threats from politicians. And, what is even more important, collecting was a necessity, because deep down nobody believed in a system that was continuously unable to provide for its citizens' basic needs for forty years or more. While leaders were accumulating words about a bright future, people were accumulating flour and sugar, jars, cups, pantyhose, old bread, corks, rope, nails, plastic bags. If the politicians had only had a chance to peek into our closets, cellars, cupboards, and drawers – looking not for forbidden books or anti-state material – they would have seen the future that was in store for their wonderful plans for communism itself. But they didn't look.

20

Epilogue

Dear Robert*: Zagreb
November 1992

In your letter you asked me if I would like to write a new afterword to my book *How We Survived Communism and Even Laughed*. I agree with you that this is a good idea, especially as I finished writing it in the spring of 1991 when the war in Croatia had not started, or at least we did not call it a war. Then everything, but everything, in Eastern Europe seemed different—more promising, I mean. Now, people are tired of waiting for a better way of life, better food, jobs and greater democracy. Moreover, they are frightened by the future, by the prospect of a continuing war that threatens to spill over the border

---

*Editor's Note: When Slavenka Drakulić was asked to write an epilogue to the paperback edition of *How We Survived Communism and Even Laughed*, she wrote the following letter to her editor as a response.

of, what was once, Yugoslavia like a contagious illness.

And this is why while I can write you this letter I cannot write a new introduction: I too am tired, what an inadequate word that sounds like. I have just finished *The Balkan Express,* a book of essays on the war. I feel totally empty, uncertain if anyone will understand what I have written about, utterly isolated and locked in by the war here in my shrunken and renamed homeland. My words seem to me like birds too young and weak to fly far.

I spent the last days of summer in the most peaceful part of the country, in the middle of the Istrian peninsula. There was almost no fighting or shelling there. When you stood on one of the hilltops overlooking a small valley with a river and road leading south, this seemed like another world entirely. But this heightened sense of tranquillity and beauty only made me feel more strongly that this war has another dimension. There is another more hidden face of war, and regardless of where we happen to be, we carry this war inside us. On the wall next to my bed I keep a postcard. It is a black and white photograph of the corner of a bedroom: The bed with white crumpled sheets seems somehow bare and naked. You can see that just before the picture was taken there were two people lying next to each other in the bed. In the untidy sheets, the mark of their bodies and the warmth they left behind, I can sense their intimacy, their love perhaps. This photograph, which was not even taken here, captures the atmosphere of loneliness that, for me, is the essence of war and I realize that there is no one I can send it to who could understand the special kind of loneliness that enters your soul in the middle of war. It is like having a piece of ice inside my chest. Where are they now? I keep asking myself. What happened to the people in this room? What happened to us? To me? To love? The emptiness, the absence of people bothers me, and makes me cry.

Once you feel the presence of death all around you cannot remain the same person; it is not only love that changes its meaning, everything alters—bread, light, water—friendship. In the last year we have had our lives changed from the outside force of war but we have internalized the war as well; our values, emotions, ethics are all different now. Several months ago in Sarajevo two babies were killed by a Serbian sniper. I wondered then what more could happen in this godforsaken blood-soaked part of the world? And immediately afterwards came the discovery of concentration camps all over Bosnia, and we understood that the extermination of Muslims and Croats meant that a new holocaust was happening here. What can happen next? Is this the end of the horror? No, I am afraid that we will have to live with this war for years. But you too will have to live with it, and it will change you—not immediately but over time. The Balkans seem so far away when you are in London or America, but that is exactly how people in Sarajevo thought when Vukovar, Osijek and Dubrovnik were shelled last winter. We all thought that the war would not come to us, that it was impossible. Wars happen to people in Latin America or in Africa, not to us in Europe. But it was not impossible, and it will take so long to understand.

You see, Robert, I still don't understand what this war really *is* even now. It took me a long time to move away from classical definitions, ideology, political concepts and the familiar media images of shelled cities and death so that I could begin to catch a glimpse of what this war encompasses.

In the beginning, war was only a word. It did not have substance for me. Journalists and political leaders used the word occasionally at first, then more frequently. But people did not utter the word because no one believed it could really happen. In the long phase of preparations and denial the substance began slowly to fill that ugly word until it became fat and real, like an insatiable dangerous animal. What was hard to realize was that the animal of war feeds only on blood. While it still seemed far away, it had a mythical quality. Everyone knew about its existence, but not many people had seen it and the stories we heard sounded so horrible and exaggerated that it was difficult to believe them. Everyone read reports, listened to the news and looked at the television images, but its mythical dimension remained preserved by the distance—the majority of us had no direct experience. In the meantime, the war came closer, the phase of denial was replaced by acceptance and adjustment. At this point, the war became real, and it was accepted as a calamity, a disaster, something that could not be prevented, a fact bigger than life.

It was a woman friend of mine who, escaping from Sarajevo and losing her apartment, her job and all her valuable possessions, became a refugee and asked me the question that now seems so important: 'What is the war?'. How strange, I thought—she is suffering the most and is asking that question. Can't she read the answer in what is happening to her? Then it occurred to me that at thirty years old she, like me or anyone born after 1945, has only lived with the idea, imagery, cliché and myths of what war might be. We had lived with the big myth of the Cold War that told us that if the two superpowers had atomic weapons World War Three would most probably be a nuclear holocaust and, as much as we found it a threat, no one believed such self-destruction was possible. Secondly, and perhaps equally important, was the idea that during the Second World War Europe had learned its lesson: There would never be another war in Europe where the worst atrocities in the history of mankind were witnessed and where the wounds have barely healed.

But now there is a war in Europe (the obvious reluctance of the West to call it war instead of *civil war,* or *a conflict* or *a tribal war,* which has much to do with the denial of war, almost as monstrous as before, is happening in Europe). This war is not nuclear but local; civilians are suffering brutalities comparable to those inflicted in the Second World War; and the largest European migration—two million refugees—of this century is taking place. Together with the end of the Cold War and the fall of Communism one is therefore forced to reexamine and redefine the concept of war itself. *'What is the war?'* The answer to this question will not be found through political analysis or historical facts; it can only be reached by understanding and articulating our relation to it.

Since the war here began my friend from Sarajevo, like me, has had time to study war; she has had a chance to read about the disastrous consequences following the formation of an artificial state in 1918, and she knows about Tito's communist state holding the nations together by force. She is well able to judge how much this war is the legacy of communism and the repression of national and religious feeling, the lack of civil society, its values and institutions. . . . This is all true, but it is not all, and does not answer her question until one is willing to see that every war begins as an external event but already lives inside us too.

This I learned when I realized that I was also reducing my friend from Sarajevo to a category—that of the refugee, and denying her personality and my responsibility altogether. In my reaction to her I recognized that my precious *I* had become *us: us* the non-refugees, *us* the 'real' citizens, *us* the Croats, and so on since once you accept the division of *us* and *them* there are endless possibilities. You see war all around you, and recognize the brutality and readiness to kill but you still keep thinking that you are different, that you are somehow better until something banal forces that righteous, knowledgeable, sensitive, intelligent person to turn to her inner self. You do not see the animal that feeds on blood, but you see clearly the seed of division, one single cancer cell from which the war multiplies and grows. Just as the cells in our organism have the ability to change into malign agents that destroy healthy tissue, so, providing that circumstances are right, a certain part of ourselves changes, eats away at our soul. It is an in-built possibility and we are responsible for it, there is nobody else to blame.

I would prefer to be writing about what happened in the countries after communism, about the fact that women were forced to be sterilized in order to get jobs in former East Germany; about the anti-abortion movement; the new autocratic leaders and the lack of democracy. I would like to have written a couple of lines about the economic crisis, or the new Berlin Walls erected at the western borders, but I cannot because doing so would mean writing about life and there is too much death around me. I wish I could have written to you about something transparent and as light as the feeling that photograph of the empty room gives me, but then I'd be sending you something else—that would be poetry. I realize that I only have words and that, from time to time, as I hold them in my arms I am less lonely.

Yours,

Slavenka Drakulić

ABOUT THE AUTHOR

Slavenka Drakulić is a respected journalist and cultural commentator in Croatia. Her work has appeared in *The Nation, The New Republic,* the *New York Times, Time,* and the *New York Review of Books* among many other publications. She was a founding member of the executive committee of the first network of Eastern European women's groups, is on the advisory boards of the Fourth International Interdisciplinary Congress of Women and *Ms.* magazine, and has received a Fulbright Fellowship for writers. Her other books include *Holograms of Fear, Marble Skin,* and *The Balkan Express.*